project
Origami

ACTIVITIES FOR EXPLORING MATHEMATICS
SECOND EDITION

project
Origami
ACTIVITIES FOR EXPLORING MATHEMATICS
SECOND EDITION

THOMAS HULL

CRC Press
Taylor & Francis Group
Boca Raton London New York

CRC Press is an imprint of the
Taylor & Francis Group, an **informa** business

AN A K PETERS BOOK

CRC Press
Taylor & Francis Group
6000 Broken Sound Parkway NW, Suite 300
Boca Raton, FL 33487-2742

© 2013 by Taylor & Francis Group, LLC
CRC Press is an imprint of Taylor & Francis Group, an Informa business

No claim to original U.S. Government works

ISBN 13: 978-1-4665-6791-7 (pbk)

Visit the Taylor & Francis Web site at
http://www.taylorandfrancis.com

and the CRC Press Web site at
http://www.crcpress.com

Dedicated to dr. sarah-marie belcastro
who brainstormed the title and concept of this book
and then supported it throughout

CONTENTS

PREFACE TO SECOND EDITION

The first edition of *Project Origami* was published in 2006, and since then I have received a lot of feedback. Every semester I get emails from people using the book in one way or another. Some are college professors or high-school teachers who tell me of an activity that they used that went well or tell me about an idea that they had or an approach that worked with their students. Some of the emails are from students themselves, asking for a pointer on a project they are attempting or asking for further resources to explore. Others still are from people who are fans of origami-math and who want to thank me for this book.

And, of course, I used this book plenty of times myself! I taught several courses on the mathematics of origami at Merrimack College and Western New England University, and whenever I teach college-level geometry, multivariable calculus, or graph theory, I draw from the activities in this book.

As any teacher knows, the act of teaching is not unidirectional, with information only passing from the teacher to the student. Rather, it is more like a feedback loop, with the teacher learning new things by watching the students learn and respond to the material. Therefore, it should come as no surprise that after years of these emails about *Project Origami* and my own teaching, new activities developed. Conversations with students and colleagues gave me ideas; sometimes a student would find an origami model themselves online or in a book and start asking mathematical questions about it. Before I knew it was happening, I discovered that I had material for half a dozen more origami-math activities. Once I realized this, I knew that producing a second edition of the book was inevitable.

This—new material generated by excitement and other people using this book— is the happy side of producing a second edition. The more embarrassing side is that in any book that contains a lot of information and gets extensively used, mistakes will be found. Many of the mistakes reported to me (or discovered by me) fit into the category of typos or regretful omissions and are easily fixed. Other mistakes, however, are of the mathematical variety. Despite the fact that the first edition manuscript went through extensive beta-testing by dozens of college and university professors (and their students) across the country, there were still mathematical errors that did not get caught.

The most egregious of these errors was in the Five Intersecting Tetrahedra activity. The solution given to that activity in the first edition was very close, but not 100% correct. This is corrected in the second edition, and in fact, a new solution is given that is much more simple than the previous one.

The process of preparing a second edition of *Project Origami* has given me the opportunity to re-read the whole book with a critical eye. I was pleasantly surprised that during the five years since the book first was published, some of my views on the simplest ways to present or teach this material have changed. Even the presentation of relatively straightforward results, like the matrix model for flat vertex folds, looked like it could be improved. Thus, nearly all of the activities from the first edition have been edited, with solutions and teaching tips improved here and there.

In my view, this second edition is a much better book that the first. The table of contents has been expanded from 22 activities to 30, over a hundred new pages have been added, and the further experiences of myself and dozen of other people (who are thanked in the acknowledgements section) have greatly improved many of the other activities. I hope you will agree that is worth it!

Thomas C. Hull
Western New England University
Springfield, MA

INTRODUCTION

Why did I write this book?

This book is the first of what I hope will be a variety of books on the mathematics of origami. It grew out of my life-long passion for the two subjects. I started learning origami at age eight, after my uncle gave me an origami instruction book. While many of the instructions in this translated-from-Japanese book were impossibly cryptic, I managed to figure out many of them and, for some reason, it stuck. Around the same time I was realizing that I was good at math, that the patterns found in addition and multiplication were easy—and fun—to memorize. I also distinctly remember noticing a link between origami and mathematics during those years. I had folded an animal, probably the classic flapping bird, and instead of putting it in my ever-growing box of folded models, I carefully unfolded it. The pattern of crease lines on the unfolded paper was intricate and lovely. Clearly, it seemed to me, there was some mathematics going on here. The pattern of lines must be following some geometric rules. But understanding these rules was well beyond my comprehension, or so I thought at the time.

My next visit to the crossroads of origami and mathematics occurred while I was in college. By that time I was well-versed in complex-level origami, having devoured the books of John Montroll, Robert Lang, Jun Maekawa, and Peter Engel. (See [Eng89], [Kas83], [Lang95], [Mon79].) I had been to a few origami conventions in New York City (hosted by the nonprofit organization now known as OrigamiUSA) and even invented a number of my own origami designs. I had also taken many math classes and was considering a career in the mathematical sciences. But then one thing happened that forced me to think about and explore the intersection of origami and math—I obtained a copy of the classic book *Origami for the Connoisseur* by Kunhiko Kasahara and Toshi Takahama [Kas87]. At first I thought this was just another complex-level origami book. In fact, I bought it because it contained instructions for John Montroll's infamous stegosaurus model (impeccably detailed and made from one uncut square of paper). Little did I know that this book also contained instructions for something that would grip my interest like a vice and result in dozens of hours of procrastination.

This book provided my first exposure to modular origami, whereby many small squares of paper are folded into identical "units" which are then locked together to form a variety of shapes. The units in *Origami for the Connoisseur* allowed one to make representations of all the Platonic solids: the tetrahedron, cube, octahedron, icosahedron, and dodecahedron. Prior to this I had only a casual understanding of these objects, but after folding many, sometimes hundreds of units to make these and other polyhedral shapes, I became intimately familiar with them. Modular origami was, quite literally, my first tutor on the subject of polyhedral geometry.

In retrospect, it is easy for me to see what was happening, although at the time I only knew that I was having fun and making beautiful, geometric objects with which to decorate my dorm room. Origami was teaching me, giving me a context in which I had to explore and master properties of various polyhedral objects. How do I arrange the units around each vertex to form a cuboctahedron? How many units of each color would I need to make an interesting coloring of the icosidodecahedron?

Over the years afterward, during graduate school and as a professor, first at Merrimack College and then Western New England University, I continued to collect everything I could find about the mathematics of origami. Since many sources were hard to find or were merely hints at underlying patterns, I often had to do my own research to help put the pieces together. In the process I saw origami intersect a variety of mathematical topics, from the more obvious realm of geometry to the fields of algebra, number theory, and combinatorics. It seemed that the more I looked, the more branches of math origami overlapped.

Simultaneous to this gathering of origami-math material, I began giving lectures on the subject for college and high-school students and their teachers. From this the interest in origami as a mathematical education tool became very clear. Teachers would regularly ask me where they could find more information on how to use origami in their classes. Eventually a few books did emerge, like [Fra99], that offered ways to use modular origami to teach geometric concepts, but none of these were done at the college level or touch the variety of topics that origami can offer.

Thus came about this book. My goal was to compile many of the origami-math aspects that I had found and present them in a way that would be easy for college or advanced high-school teachers to use in their classes.

How to use this book

Thirty activities are included in this book that cover a variety of mathematical areas. The intent is for mathematics instructors to be able to find something of use no matter what college or, perhaps, high-school course is being taught.

Each activity begins with a list of courses in which it might fit. In the appendix you'll find a cross-reference that lists which activity might match a given course.

However, it is important to realize that many of these activities can be effectively used in a variety of courses at a variety of levels. The angle trisection activity, for example, has been very popular among high-school geometry teachers, yet it also makes for a very illustrative diversion in an upper-level Galois Theory course. The PHiZZ unit Buckyball activity can be a great extended project for "liberal arts" general-education math classes, but it also provides a hands-on way for students of graph theory to explore the connections between 3-edge-coloring cubic planar graphs and the Four Color Theorem, not to mention the opportunity to classify geodesic spheres in an upper-level geometry class.

In short, the key word in terms of using this book is *flexibility*. Each activity includes handouts that may be photocopied and used with your class as well as notes to the instructor on solutions, how the handouts can be used, suggestions on pedagogy, and further directions that can be taken.

But you, depending on your class, your time, and your interest in origami, can and are encouraged to find your own ways to use this material. Perhaps it would be better for you to present the Folding Equilateral Triangles in a Square and Can Origami Trisect an Angle? activities at the same time. Perhaps you'd prefer to use only part of a handout or add on investigative questions of your own. Perhaps these would fit your class better as homework or extra-credit assignments. Perhaps one of these activities could be the basis for a senior research project. Perhaps you could spend a whole year using these activities for your college's Math Club or your high school's Math Circle.

To provide readers with as much flexibility as possible, our publisher, A K Peters/CRC Press, is making all of the handouts in this book available online. Many professors who beta-tested these activities had access to PDF versions of the handouts and thus were able to modify them to taste. Some copied the graphics into separate documents so that they could write their own text and modify the questions. Others removed certain questions or combined several activities into one. Others chose to insert more explanation for their students. You are the one most familiar with your students, and thus we want to give you the ability to tailor the activities to your liking.

The online PDF versions of the handouts can be found on the CRC Press web site: go to http://www.crcpress.com/product/isbn9781466567917 and click on the "Downloads" tab.

I am especially interested to know what people do with these activities. If you modify or find interesting ways to utilize them, please feel free to email me and share your experiences: thull@wne.edu.

Discovery-based learning

The main pedagogical approach behind all the activities is one that is active and discovery-based (as opposed to, say, a lecture-based approach). There is a logical choice for this that deserves some explanation.

One of the main attractions of using origami to teach math is that it requires hands-on participation. There's no chance of someone hiding in the back of the room or falling asleep when everyone is trying to fold a hyperbolic paraboloid (see the Rigid Folds 1 activity). The fact that origami is, by definition, hands-on makes it a natural fit for active learning. One could even make the argument that while folding paper, especially when making geometric models, latent mathematical learning will always happen. There's no way a student can make a dodecahedron out of thirty PHiZZ units without an understanding of some fundamental properties of this object.

Therefore, when choosing to use origami as a vehicle for more organized mathematics instruction, an easy choice is to let the students *discover things for themselves*. This approach to teaching mathematics, where students are allowed to experiment and discover basic principles and theorems themselves, was pioneered by David Henderson in college-level geometry courses (see [Hen01]). The approach is based not only on exploration but also in students *learning how to ask the right questions* while exploring.

I tried to achieve a mixture of this in the handouts in this book. Some of them try to lead the student toward asking the proper questions that lead to theorems, like in the Haga's "Origamics" activities. Others, like the Exploring Flat Vertex Folds activity, is deliberately very open-ended. The specific purpose in such open-ended activities is for students to gain experience with asking questions and building conjectures.

I highly encourage instructors to not shy away from this approach. Too often professors feel that they need to instruct their students on the fine art of conjecture-building. But the best way to learn this process is to just do it. Some students behave as if they were just waiting to be asked to make conjectures; once you get them going, they can't stop! Others do have difficulty with open-ended assignments, but again, these difficulties arise from not knowing how to ask questions. Engaging such students in a Socratic dialog often helps a lot.

For example, a student who can't find any patterns in flat vertex folds might be asked, "Well, is there anything going on with the mountain and valley creases?" If that doesn't help, then a more specific question, "How many of them are there at your vertex?" will get things going. It's these questions that help students see that the piece of folded paper is their experimental laboratory. The math ceases to be an abstract entity, only existing in their mind. It becomes tangible, something they can hold in their hand and count or use to compute data from which patterns, conjectures, and theorems flow.

Nothing gives students a feeling of ownership of such discovery like the personal touch of their own names. Sure, the fact that the difference between the number of mountain and valley creases at a flat foldable vertex is always two is known as Maekawa's Theorem (see the Exploring Flat Vertex Folds activity). But it might as well be christened "Danielle's Conjecture" for a few classes as students discover and try to prove it.

However, it should be noted that a completely 100% discovery-based instruction method is not for everyone. Instructors who are more comfortable with lecture-based instruction can still use the handouts for, say, 20 minutes of in-class activity and then wrap up the main points and student observations via lecture. Still, it might be more interesting to see what the students have come up with and ask some of them to present their results to the whole class.

The value of the discovery-based approach should be clear, in that it provides students with the experience of being a mathematical researcher. If helping train your students for independent research or for a senior capstone experience is one of your goals, then by all means give this approach a try. In fact, if you think about the skills and experiences needed to become comfortable with mathematical inquiry, you just might end up totally changing the way you approach teaching your course. For example, it's very important for math researchers to understand that not succeeding is OK—that failure is a natural part of discovery. Thus, when exploring one of these origami activities in class, instructors should be prepared for their students to *not succeed* and realize that this is fine. This leads to the next topic.

Preparing yourself

The preparation required for these activities takes several forms.

First of all, if the activity has a strong folding component, like folding modular units or folding a crane, instructors need to practice folding these things themselves in advance. What's more, instructors need to think, as they fold, about how they would explain the folding process to a classroom of students or to individuals who are stuck. Teaching origami is quite a bit different from teaching math. It involves trying to communicate three-dimensional movements by "show and tell." The handout instructions for folding in these activities are meant to help, but some people have a very hard time translating two-dimensional instructions into three-dimensional movements of their hands and paper. *Always assume* that there will be students who need one-on-one help with the folding instructions.

If the technology is available, using a document camera (also known as a digital imager or Elmo) can be a big help. Document cameras allow one to place their hands and a piece of paper underneath a camera that will then project this image on an overhead screen. Using this, a whole class can see what your hands are doing, up-close, as you fold the paper. In my experience, this is by far the most efficient way of teaching a whole class to fold paper. It also works very well for showing how to lock modular units together. Such units are often small, and a good document camera will allow you to zoom-in on the details of putting the units together properly.

Note, however, that while it is important for instructors who, say, are using the Making Origami Buckyballs activity to become very familiar with the PHiZZ unit, understand its locking process well, and make a 30-unit dodecahedron of their own (and properly 3-color it), other longer projects can be left to the students to figure out. Instructors are not likely to have the time to make a 270-unit Buckyball or an 84-unit torus beforehand, although these projects are fun and make great office decorations. Students should be encouraged to attempt larger projects. The fact that you might not have done them yourself can give students an extra feeling of accomplishment over their achievements.

Aside from the paper folding itself, it goes without saying that all instructors will have to tailor these activities to their own classes. The chances of a successful experience with these activities will increase dramatically if you make sure that your goals and expectations of the activity are clearly focused. Is your main goal to reinforce student understanding of Euler's formula and its uses (as in the Buckyball activity)? Is it for your students to see hands-on applications of the algebra of $\mathbb{Z}_n$ and number theory (as done in the Folding Strips into Knots and Fujimoto Approximation activities)? Or do you see the main goal as being to introduce more active participation in class or for students to explore and discover mathematics on their own?

The answers to these questions will allow you to clarify how to use the activity in class—how much time to spend on the hands-on part versus group discussion, or whether to assign the folding instructions for homework beforehand, or whether to expect the students to come up with very many conjectures on their own in class. Of course, the first time any of us try a new activity, especially one with an active or discovery-based learning component, it needs to be thought of as an experiment. The second time you try using any of these activities will require much less preparation.

Where to find paper

The question of where to obtain paper is a bit complicated. It entirely depends on what you or your students will be folding. While paper is paper, it comes in many different types. Some projects and activities can be done with any type of paper, but often there are preferences that can make the students' and instructor's job easier. I'll break these preferences down into categories.

For PHiZZ units, Flat Vertex Fold, Haga's Origamics, Matrix Model, Butterfly Bomb activities

I recommend three-inch square memo cube paper, which can be easily bought (for about $3 for 500 sheets) from office supply stores and comes in a rainbow of colors. Look for it near the Post-it note section, but *make sure* you do not buy Post-it notes! (The sticky side gets in the way of folding and sliding modular units into one another.) The best memo cube paper is the type that comes in its own plastic container—this paper is more accurately square than the type that doesn't come in a box. Also, if you look carefully you can find *blank* memo cube paper. If you're unlucky all they'll have is paper that's blank on one side and has "while you were out" office messages printed on the other side. That works just as well, and your students may find it more humorous anyway.

Business cards

Once you get bitten by the business card modular origami bug (and yes, there are many other modular units to be made from business cards than those presented in this book), you'll be very interested in collecting large supplies of discarded cards. This can sometimes be very easy to do. Visit an office supply or printing store where they print business cards for customers and ask if they have any unwanted cards. Often such places will have boxes of cards with printer errors or that were never picked up and have been sitting around for months. If you make it known that you're interested in such unwanted business cards, they'll often save such prizes for you when they turn up.

In a pinch, you can buy blank cards, but ones with printing on them can be much more interesting. Along those lines, be on the lookout for colorful or nicely patterned business cards at restaurants. Pinching ten or so of these at a time can slowly build a good collection. You can also ask students to acquire business cards of their own beforehand and bring them to class.

Strips of paper (for folding knots)

It can be difficult to find rolls of thin paper. Ticker-tape paper is ideal, but you can also get rolls of accounting tape, which is the paper accountants use for those calculators that print out the calculations as they go. You can usually find rolls of such paper at office supply stores.

Actual origami paper

This is paper that is colored on one side and white on the other, and origamists often call it "kami" or "plain kami." It folds very well and is considered "special" origami paper. It is the paper you probably want students using if they are folding cranes (for the Folding and Coloring a Crane activity) or other traditional origami models. You can find it at any art supply store. It usually costs $5–$6 (US) for 100 squares, 6 inches per side, in a variety of colors. You can also order it on the web at OrigamiUSA (a national nonprofit organization—if you're an advocate of origami, or want to become one, you should become a member, since it gives you a magazine, access to members-only web content, the ability to attend origami conventions, and a 10% discount on buying things from them. See http://origamiusa.org).

Other options (and the Five Intersecting Tetrahedra)

The most basic paper you can use is photocopy paper. You can use up that pile of scrap 8 1/2 inch by 11 inch paper you have stacked in your office by cutting it into squares with a paper cutter. This makes great all-purpose, no frills paper to use in class. It's fine paper to use when folding cranes and is *very* good to use when making the Five Intersecting Tetrahedra model, since it is heavier than normal origami paper. Also, you can get it in a variety of colors from any office supply store or from the Print Center at your college or university.

In fact, a very good resource for square paper and business cards (and maybe even strips of paper) is your friendly Print Center on campus. While not everyone has a friendly Print Center at their school, it would be worth your time to find out if you do. Pay them a visit and tell them that you're doing origami in your classes. They'll probably be happy to cut paper to size for you or give you discarded business cards, or they might have long strips of paper handy.

Other sources

In each activity I've tried to provide references for the material as well as for places where more information can be learned.

Since interest in origami-math has been increasing, there are some books now available that are devoted to certain aspects of the subject. Also, there are a few books with chapters devoted to paper folding as well as some proceedings and other books that are useful. Since these sources might be very valuable, depending on your specific interests in origami-math, they deserve special mention.

Galois Theory by David Cox [Cox04]. This book is excellent anyway because David Cox is such a good writer. But Chapter 10 is devoted to geometric constructions, and Section 3 of this chapter is on origami. This is probably the best exposition of an algebraic, Galois Theory approach to origami geometric constructions available. Instructors interested in using the Folding a Parabola, Can Origami Trisect an Angle?, and Solving Cubic Equations activities in an advanced algebra class should consult this book.

Geometric Folding Algorithms: Linkages, Origami, Polyhedra by Erik Demaine and Joseph O'Rourke [Dem07]. This book is a must-read for anyone interested in the field of computational origami, where questions are asked about how feasible it is to find answers to thorny, and even not-so-thorny, folding problems. As the title indicates, the authors look at folding and unfolding linkages (which can be thought of as one-dimensional folding), paper, and polyhedra. Anyone who is confused about why this would be an active area of research in theoretical computer science should look at this book and see the applications to robotics, protein folding, and numerous other things. Much of the math presented in this book is in alignment with the math presented in *Project Origami*.

Geometric Origami by Robert Geretschläger [Ger08]. This book focuses on origami geometric constructions, giving a very axiomatic and synthetic approach. Fans of geometry and the Folding a Parabola and angle trisection activities will enjoy Geretschläger's book very much.

Origamics: Mathematical Explorations Through Paper Folding by Kazuo Haga [Haga08]. This book is an English translation of many of Haga's Japanese writings on using very simple, geometric folding problems to engage students in the process of mathematical discovery. Haga's approach to geometric origami is rather unique, and you can get a big taste of it in the Haga's "Origamics" activity. If you like that, then definitely buy this book.

Origami^3 [Hull02-2], *Origami*^{4} [Lang09], and *Origami*^{5} [Wang11]. These three books are the proceedings of the third, fourth, and fifth international meetings of Origami Science, Mathematics, and Education (OSME, for short). The first two such meetings took place in Italy (1989) and Japan (1994), but the proceedings for those meetings are out of print and very hard to find. The other proceedings are still in print and present excellent snapshots of the state of origami research in science, math, and education. While I am, as editor of one of the volumes, biased, I feel confident in saying that no matter what your taste in origami you'll find many of the articles in these books of great interest. Make sure your library has them if you would like your students to be able to see what current research in origami is like.

Origami Design Secrets: Mathematical Methods for an Ancient Art, Second Edition, by Robert Lang [Lang11]. Robert Lang is one of the pre-eminent creators of complex, artistic origami models, and this book is his *magnum opus*. It describes in detail Lang's *TreeMaker* algorithm as well as other origami design techniques. While none of the activities in *Project Origami* deal directly with origami model design (that is, trying to answer the question, "How do you fold an insect from a square without making any cuts?"), the techniques that modern origamists use follow from mathematical principles of origami (for example, things like Maekawa and Kawasaki's Theorems from the Exploring Flat Vertex Folds activity). Students who get bitten by the origami bug should devour this book. It's a great source for student projects in this area.

Mathematical Reflections: In a Room with Many Mirrors by Peter Hilton, Derek Holton, and Jean Pedersen [Hil97]. This book (in Springer's Undergraduate Texts in Mathematics series) has a 57-page chapter titled Paper-Folding and Number Theory. It collects much of the research done by Peter Hilton and Jean Pedersen on the number theory behind folding strips of paper into polygons and polyhedra. This is very related to the topics covered in my Fujimoto Approximation and Folding Strips into Knots activities, although Hilton et al. use a different approach and take the material in different directions. If these activities appeal to you, definitely explore this chapter.

Origami for the Connoisseur by Kunihiko Kasahara and Toshie Takahama [Kas87]. Of the many origami instructions books in print, this one is the most mathematical (and was mentioned earlier in this Introduction). It contains instructions for many geometric models, like polyhedra and spiral shells, both from single sheets of paper and modular. It also contains references to Maekawa and Kawasaki's Theorems as well as some of Haga's origamics activities. While several of the models are very complicated, requiring expert origami skills, others are surprisingly simple and elegant. This is a gem of a book.

Geometric Constructions by George E. Martin [Mar98]. The last chapter (14 pages) of this book is devoted to geometric constructions via paper folding. Martin's approach is purely geometric, as opposed to Cox's algebraic analysis, so this would appeal to teachers of geometry who want to learn more about origami geometry. Martin concentrates on only the most sophisticated of the single-fold origami operations—the one explored in the Solving Cubic Equations activity. This is all one needs, however, to perform constructions such as angle trisections and cube doublings. Martin also compares this to other construction methods, for instance, using a marked ruler.

Origami Polyhedra Design by John Montroll [Mon09]. John Montroll is an origami legend. He was one of the first people to achieve the level of complexity in origami (as seen in his 1979 book [Mon79]) that we today associate with complex-level origami. He also is very interested in folding polyhedral shapes from single, uncut squares, and in this book he provides instructions for many such folding projects as well as explains the math behind them. In this way he shows some great applications of trigonometry as well as planar and 3D geometry, making the math used very accessible for motivated high-school students. Teachers looking for interesting ways to show how trigonometry and geometry are used in origami would enjoy this book.

How to Fold It: The Mathematics of Linkages, Origami, and Polyhedra by Joseph O'Rourke [ORo11]. As the title suggests, this book is thematically similar to Demaine and O'Rourke's book [Dem07], but at the same time it is very different! O'Rourke's *How to Fold It* is much smaller and meant for high-school or college students to follow (whereas [Dem07] is more of a research monograph). The book contains many projects with clear explanations and would make a good companion text for *Project Origami*.

Fragments of Infinity by Ivars Peterson [Pet01]. This is a popular math book for a general audience and has a 22-page chapter on origami called Plane Folds. While not a math text, it does give a good overview of flat origami crease patterns, Maekawa's Theorem, Lang's *TreeMaker* algorithm, and origami tessellations. In particular, it includes some wonderful pictures of Chris Palmer's complex folded tessellations. If you found the Folding a Square Twist activity exciting, definitely check this out.

Geometric Exercises in Paper Folding by T. Sundra Row [Row66]. This book is a classic. T. Sundra Row was an Indian mathematics teacher who, in the late 1800s, wrote this book on the basic geometric constructions that can be performed by paper folding. It attracted the attention of Felix Klein, and after he referenced it in some of his publications, Western publishers began printing it world-wide. The latest printing was by Dover, and it should not be hard to find in most libraries. A careful reading of the book makes it unclear whether Row knew that origami could do things like trisect angles (no method for this is given in the book, but Row does discuss how paper folding relates to solving some types of cubic equations). Nonetheless, this is an excellent source of methods for folding a variety of polygons and shapes in paper. While written in the very formal style of over a hundred years ago, the construction methods are simple and could easily be adapted for modern geometry classes (for both college and high school).

ACKNOWLEDGMENTS

This book was made possible via a variety of support. First and foremost is the Paul E. Murray Fellowship that I received at Merrimack College which funded the creation of the book's first draft. Without the generosity of the Murray family, this project might never have gotten off the ground. In general both Merrimack College and the Hampshire College Summer Studies in Mathematics have been incredibly supportive by providing me with environments in which to study and teach the mathematics of paper folding freely and creatively.

I have been very lucky to receive the help and input from a large number of people in the creation of this book. Discussions and feedback from those in the origami and mathematics community have been invaluable: Roger Alperin, sarah-marie belcastro, Ethan Berkove, Vera Cherepinsky, David Cox, Erik and Marty Demaine, Koshiro Hatori, Miyuki Kawamura, David Kelly, Jason Ku, Robert Lang, Jeannine Mosely, James Tanton, Tamara Veenstra, and Carolyn Yackel.

I was also fortunate to have the help of many Project NExT fellows who beta-tested these activities in their own mathematics classes during the spring and fall of 2005. Project NExT (which stands for New Experiences in Teaching) is a fellowship program of the Mathematical Association of America designed to help new math professors become better teachers and scholars without becoming lost or overwhelmed by the academic mathematics community. Their collective pedagogical wisdom and experience has directly shaped this book. Furthermore, through Project NExT word of this book spread into the greater mathematical community, where numerous other faculty and students in graduate school, college, and even high-school asked to be beta-testers. In particular, I need to thank Cristina Bacuta, Don Barkauskas, Mark Bollman, David Brenner, Kyle Calderhead, Scott Dillery, Melissa Giardina, Susan Goldstine, Aparna Higgins, Barbara Kaiser, Michael Lang, Chloe Mandell, Hope McIlwain, Blake Mellor, Andrew Miller, Cheryl Chute Miller, Donna Molinek, George Moss, Katarzyna Potocka, Jason Ribando, Liz Robertson, Cameron Sawyer, Amanda Serenevy, Brigitte Servatius (and her students Roger Burns, Onalie Sotak, and John Temple), Linda Van Niewaal, Kathryn Weld, Jennifer Wilson, and Yi Zhou.

Thanks must also go to all the students that I've had in my classes where origami-math has been taught, which has included the University of Rhode Island, Merrimack College, the Hampshire College Summer Studies in Mathematics, the University of Cincinnati, and Western New England University. Not many people realize, I think, that if you are a teacher who cares about what you do and thinks deeply about it, you learn just as much from your students as they learn from you. The students that I have learned from are too numerous to mention, but I would like to thank Hannah Alpert, Mike Borowczak, Michael Calderbank, Alessandra Fiorenza, Emily Gingras, Josh Greene, Monique Landry, Kevin Malarkey, Wing

Mui, Emily Peters, Gowri Ramachandran, Jan Siwanowicz, Ari Turner, Jeanna Volpe, Haobin Yu, the 1995–1996 graduate students in mathematics at the University of Rhode Island, and my Spring 2005 Combinatorial Geometry class at Merrimack College, who were unknowing guinea pigs for these activities as the first edition of book was being finished.

Also, many of these activities were tested on and shared with the middle- and high-school teachers who took my Origami in Mathematics and Education course as part of Western New England University's Master of Arts for Teachers in Mathematics (MAMT) program. These "students" would learn some origami-math in my class and then the next day use it in their own classes, giving me immediate feedback. In particular, ideas generated by Ann Farnham and Diane Glettenberg influenced some of the new activities in the second edition of *Project Origami*.

Activity 1
FOLDING EQUILATERAL TRIANGLES IN A SQUARE

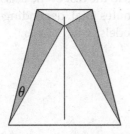 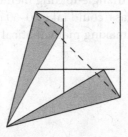 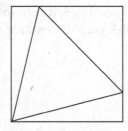

For courses: precalculus, elementary algebra, trigonometry, geometry, calculus (optimization), modeling

Summary

Students are asked to find a way to fold an equilateral triangle from a square piece of paper. Then the challenge of finding the largest possible equilateral triangle that can be folded from a square is given. Of course, students need to prove that their conjectured triangle is the largest possible.

Content

The geometry component of this problem only requires the ability to work with 30°-60°-90° triangles. However, more creative geometrical insights can lead to more elegant solutions.

For a calculus class, this problem could actually be posed without any mention of origami: What is the largest equilateral triangle that can be inscribed in a square? But knowing that paper folders actually use this knowledge can provide extra motivation. This is a challenging modeling problem that can be completely done without resorting to derivatives, provided the students set up the model carefully, know trigonometry solidly, and do a proper graphical analysis. As an optimization problem, it breaks away from the mold that is typically encountered in calculus textbooks, thus forcing students to apply their knowledge to a brand-new, real-life situation.

Handouts

Three optional handouts are provided:

(1) Introduces the general problem of folding an equilateral triangle inside a square.

1

(2) Provides a few guided steps in setting up the optimization model.

(3) Leads students step-by-step through the optimization model.

Time commitment

Handout 1 will require about 40 minutes of class time, including student exploration and presentation of their triangle-folding methods to the rest of the class.

Handout 2 or 3, if done in class, could take 50–60 minutes total, depending on how quick your students are at making mathematical models.

How to Fold an Equilateral Triangle

The goal of this activity is to fold an equilateral triangle from a square piece of paper.

Question 1: First fold your square to produce a 30°-60°-90° triangle inside it. Hint: You want your folds to make the hypothenuse twice as long as one of the sides. Keep trying! Explain why your method works in the space below.

Question 2: Now use what you did in Question 1 to fold an equilateral triangle inside a square.

Follow-up: If the side length of your original square is 1, what is the length of a side of your equilateral triangle? Would it be possible to make the triangle's side length bigger?

What's the Biggest Equilateral Triangle in a Square?

If we are going to turn a square piece of paper into an equilateral triangle, we'd like to make the **biggest possible** triangle. In this activity your task is to make a mathematical model to find the equilateral triangle with the **maximum area** that we can fit inside a square. Follow the steps below to help set up the model.

Question 1: If such a triangle is maximal, then can we assume that one of its corners will coincide with a corner of the square? Why?

Question 2: Assuming Question 1, draw a picture of what your triangle-in-the-square might look like, where the "common corner" of the triangle and square is in the lower left. Now you'll need to create your model by introducing some variables. What might they be? (Hint: One will be the angle between the bottom of the square and the bottom of the triangle. Call this one θ.)

Question 3: One of your variables will be your *parameter* that you'll change until you get the maximum area of the triangle. Pick one variable (and try to pick wisely—a bad choice may make the problem harder) and then come up with a formula for the area of the triangle in terms of your variable.

Question 4: With your formula in hand, use techniques you know to find the value of your variable that gives you the maximum area for the equilateral triangle. Be sure to pay attention to the proper range of your parameter.

Question 5: So, what is your answer? What triangle gives the biggest area? Find a folding method that produces this triangle.

Follow-up: Your answer to Question 5 can also give a way to fold the largest *regular hexagon* inside a square piece of paper. Can you see how this would work?

What's the Biggest Equilateral Triangle in a Square?

In this activity your task is to find the biggest equilateral triangle that can fit inside a square of side length 1. (Note: An equilateral triangle is the triangle with all sides of equal length and all three angles measuring 60°.) The step-by-step procedure will help you find a mathematical model for this problem, and then to solve the optimization problem of finding the triangle's position and maximum area.

Here are some random examples:

Question 1: If such a triangle is maximal, then can we assume that one of its corners will coincide with a corner of the square? (Hint: The answer is yes. Explain why.)

Question 2: Assuming Step 1 above, draw a picture of what your triangle-in-the-square might look like, where the common corner of the two figures is in the lower left. (Hint: See one of the four examples above.) Now you'll need to create your model by labeling your picture with some variables. (Hint: Let θ be the angle between the bottom of the square and the bottom of the triangle. Let x be the side length of the triangle.)

Question 3: Come up with the formula for the area of the triangle in terms of one variable, x. Then, find an equation that relates your two variables, x and θ. Combine the two to get the formula for the area of the triangle in terms of only one variable, θ. (Hint: Your last formula will be $A = \frac{\sqrt{3}}{4}\sec^2\theta$.)

Question 4: What is the range of your variable θ? Explain. (Hint: The range should be $0° \le \theta \le 15°$.)

Question 5: Most important part: With your formula and the range for θ in hand, use techniques of optimization to find the value of θ that gives you the maximum area for the equilateral triangle. Also, find the value of this maximum area. (Hint: For simplicity, you may want to express all trigonometric functions in terms of sin and cos).

SOLUTION AND PEDAGOGY

Folding an equilateral triangle

There are a number of ways to fold an equilateral triangle in a square. All involve finding a way to produce a 60° angle. Your students might find new and creative ways to do this, but the most common way people discover is shown below. (We assume in these pictures that the side of the original square has length 1.)

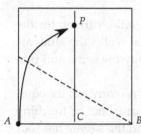 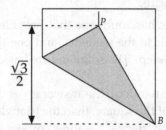 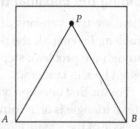

The origami "move" here is to take one corner, A, and fold it to the center line (so the paper must have been creased in half first) *while at the same time* making sure the crease you make goes through corner B.[1] We let P be the image of point A under this fold. Then we have that ABP is an equilateral triangle. This can be seen in a number of different ways:

- Let C be the midpoint of AB. Then considering $\triangle BCP$, we have that BP has length 1 (since it is the image of AB) and BC has length $1/2$. The Pythagorean Theorem then tells us that CP has length $\sqrt{3}/2$, so $\triangle BCP$ is a 30°-60°-90° triangle. Then creasing AP gives us an equilateral triangle.

- Since BP is the image of AB under the folding, BP has length 1. We can then either say, "Now fold B to the center line in the same way," or "By symmetry," to get that AP has length 1 as well. Thus $\triangle ABP$ is an equilateral triangle.

In the solution pictured here, the length of the side of our triangle is the same as the side of the square. However, if we imagine rotating the triangle counterclockwise a little bit about the point A, we could then expand the sides some and still remain inside the square. So it *is* possible to make a bigger equilateral triangle inside the square.

Pedagogy. Many students will first try to construct a 30°-60°-90° triangle by trying to make the right angle be at a corner of the square. This is not the easiest thing to do, and suggesting that such students try folding the corner inside the square instead can get them over this mental block. Suggesting that they use the $1/2$ center line can also be offered.

---

[1] This is a standard origami move: A point p_1 is folded onto a line l, but that is not enough to determine where exactly the crease should be made. So a second point, p_2, is needed, where we make sure the crease line goes through p_2 as well as making p_1 land on l. See the Folding a Parabola activity for more information.

Oftentimes students overhear ideas from other groups in class, or a good idea gets suggested from one group to another. That's fine, but everyone should write down a proof that their triangle is really 30°-60°-90° or equilateral. Groups should present their proofs to the class so that everyone can see that it can be done in more than one way. Writing up their proofs formally can be assigned individually for homework, if desired. (This should be easy after the group work, but writing things up "for real" is still a very valuable activity.)

Finding the maximal triangle

There are two versions of this handout: one that provides only a frame for the problem, leaving all the details to the students, and one that walks the students through the problem, step-by-step. The solutions are basically the same and presented here in tandem.

For the first question on the handout, the answer is yes. If no corner of the equilateral triangle is on a corner of the square, then the triangle must not be touching one side of the square (since the triangle has three corners and the square has four sides). Assume this is the left side. Then the three corners of the triangle must be touching the three other sides of the square, for otherwise we could make the triangle bigger. Then we can slide the triangle to the left until it touches this left side with one of the corners that touches either the top or bottom side as well. This puts a corner of the triangle on a corner of the square.

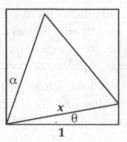

To set up the model, students will need a picture something like the above figure. The base of the triangle (length x) should extend from the bottom left corner to the right side of the square. Then we need to consider the range $0° \leq \theta \leq 15°$, for if $\theta > 15°$, then we'll have $\alpha \leq 15°$ and we'd be in a case symmetric to one with $\theta \geq 15°$. In other words, the symmetry of the square restricts the range of θ that we need to consider.

We need to find a formula for the area A of the equilateral triangle and then try to maximize this formula in terms of θ. (We want to do this in terms of θ, instead of x, because θ is the variable that tells us the position of the triangle in the square.) Since the base of the triangle is x, its height is $(\sqrt{3}/2)x$. So $A = (\sqrt{3}/4)x^2$, but we wanted it in terms of θ. Well, $\cos \theta = 1/x$, so $x = 1/\cos \theta = \sec \theta$. Thus we have

$$A = \frac{\sqrt{3}}{4} \sec^2 \theta.$$

We could take the derivative of this and try to maximize it using calculus, but we don't really need to. Since $\cos\theta$ is a decreasing function on the interval $0 \leq \theta \leq \pi/12$ (we really should be working in radians, after all), we know that $\sec\theta$ is an increasing function on this interval. The same will be true of $\sec^2\theta$, so the maximum value of A will be on the right-most endpoint of the interval, $\theta = \pi/12$. Students can see this by graphing the function $A(\theta)$:

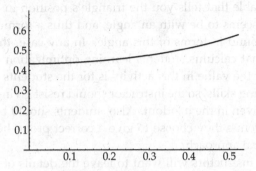

Thus the maximum area is achieved at $\theta = \pi/12 = 15°$. This results in a picture where one corner of the triangle is on a corner of the square and the triangle is symmetric about a diagonal of the square.

Students who do use derivatives to solve this would get

$$\frac{dA}{d\theta} = 2\frac{\sqrt{3}}{4}\sec^2\theta\tan\theta = \frac{\sqrt{3}\sin\theta}{2\cos^3\theta}.$$

Since $0 \leq \theta \leq 15°$, we know that $dA/d\theta = 0$ only when $\theta = 0$. This means that the area formula has a critical point at $\theta = 0$. But this is just an endpoint of our interval, so this means that the extreme values of the area A will happen at the endpoints $\theta = 0$ and $\theta = 15°$ (since there are no critical points in between). The question then is, which is a maximum and which is a minimum? We could take the second derivative of A and determine the concavity of the critical point $\theta = 0$, but taking such a derivative looks a little foreboding. Instead we could just check the value of A when $\theta = 0°$ and $\theta = 15°$. Fifteen degrees wins.

Students who do both of these handouts should be able to find a folding sequence for the maximal equilateral triangle. The pictures below serve as such a folding sequence as well as a "proof without words" that it works. (First note that $\theta = 15°$ in the left-most figure.) This folding sequence proof was developed by Emily Gingras, Merrimack College class of 2002.

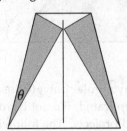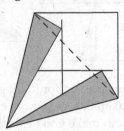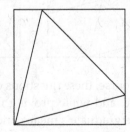

Pedagogy. Students familiar with the classic "fenced-in pen along a side of a barn" or "box folded out of a sheet of cardboard" calculus problems should see right away that our maximum equilateral triangle problem should be solvable using similar methods. However, the model for our problem is very different from those classic ones, and most students find it very challenging to set up the model properly. The hard and subtle part is making sure that you can parameterize the problem with a variable that tells you the triangle's position in the square. The best way to do this seems to be with an angle, and thus a formula for the triangle's area must be found in terms of this angle. In any case, this problem is at the right level of what calculus students learning optimization problems *should* be able to solve. But the value in this activity is for the students to sharpen their mathematical modeling skills, so the instructor should resist giving any more hints than those already given in the handout. Also, students should be encouraged to explore whatever avenue they choose to give a correct proof, be it a numerical, graphical, or analytical approach.

However, not all instructors will want to leave the details of such an activity entirely open. The second version of the optimization handout is for those who would like their students to see the proper procedure for such a problem and work out the details themselves. The format and pacing of this handout follows a suggestion by beta-tester Katarzyna Potocka of Ramapo College of New Jersey.

It can also be valuable to do this activity in a geometry course to emphasize the interconnections between mathematical disciplines. Typically, math major undergraduates in an upper-level geometry course will claim to have forgotten all of calculus, making this all the more worthy to do.

Follow-up activity

If you think about how a maximal regular hexagon would be inscribed in a square, as in the pictures below, and make horizontal and vertical half-way creases, you can see that one quarter of the square is exactly like the crease pattern for the maximal equilateral triangle. Therefore, the folding method for the triangle can be modified to give a maximal hexagon. The far right figure below abbreviates such a method.

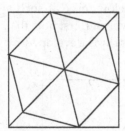 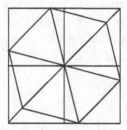 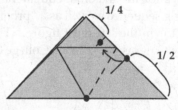

Of course, these questions can be asked for folding any regular polygon inside a square, and while proving maximality gets more complicated, it's not beyond an undergraduate's means and can make good extended projects. The following figures show a way of proving the maximal hexagon case. Let θ be the angle it makes with the bottom edge of the square (whose side length is, again, 1) and let x be the length of a side of the hexagon. The hexagon is made up of six equilateral triangles, which makes it easy to compute the area of the hexagon: $A = 6 \times$ (area of one triangle) $= 6(x/2)(\sqrt{3}/2)x = (3\sqrt{3}/2)x^2$. But we want to maximize this with respect to θ.

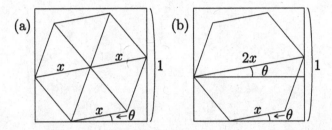

Figure (b) shows how we can do this. The diameter of the hexagon is $2x$, and if we assume that two opposite corners of the hexagon will be touching the left and right sides of the square, then we can form a right triangle from one of these corners (the left one, in the figure) with base of length 1 and hypothenuse one of the diagonals (length $2x$). Since the bottom of this triangle is parallel to the bottom of the square, and hypothenuse is parallel to the bottom of the hexagon, we know that the base angle in this right triangle will be θ. Thus $\cos\theta = 1/2x$, or $x = (1/2)\sec\theta$. Thus the area of the hexagon is $A = (3\sqrt{3}/8)\sec^2\theta$.

To maximize this, we need to find the range of θ we need to consider. The symmetry of the hexagon shows us that $0° \le \theta \le 15°$ is all we need to consider. Like the triangle case, the largest endpoint of this interval, $\theta = 15°$, gives the largest area. This will make one of the diagonals of the hexagon lie along a diagonal of the square.

Activity 2
ORIGAMI TRIGONOMETRY

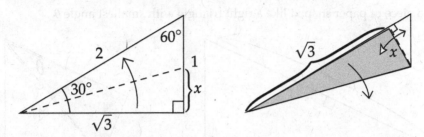

For courses: trigonometry, precalculus

Summary

Paper right triangles are folded to explore some advanced trigonometry ideas.

Content

Students are led through a paper-folding method to prove the double-angle formulas for $\sin x$ and $\cos x$. A different folding of $45°$-$45°$-$90°$ and $30°$-$60°$-$90°$ triangles is used to discover exact expressions for the side lengths of a $15°$-$75°$-$90°$ triangle and a $22.5°$-$67.5°$-$90°$ triangle, respectively. From this, exact expressions for sine, cosine, and tangent can be found for $15°$, $75°$, $22.5°$, and $67.5°$ angles.

Both of these activities employ basic knowledge of sine, cosine, and tangent (for example, knowing sine is opposite over hypothenuse). The second activity uses similar triangles in one of the steps as well as the ability to manipulate square roots algebraically (for example, multiplying the top and bottom of a fraction by the conjugate radical).

Handouts

There are two handouts, one for each activity described above.

Time commitment

If pre-cut right triangles are made available, then the folding part of these activities is quite trivial. The majority of the time commitment is for students to answer the questions on the handout and do any necessary algebra on paper. Planning on at least 30 minutes for each handout is probably safe, but a lot will depend on how comfortable the students are with trigonometric functions and, for the second activity, square roots.

Proving the Double-Angle Formulas

Make a piece of paper shaped like a right triangle with smallest angle θ.

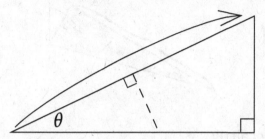 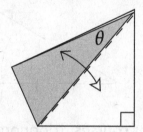

Fold the corner of the smallest angle to the other corner, as shown above. Then fold along the edge of the flap that you just made. Unfold everything.

The result of your folding will be that your triangle is divided into three smaller triangles, as shown to the right. Label the points of this figure A through D and O, as shown, and let $AO = 1$ and $OC = 1$. (You can think of O as the center of a circle of radius 1.)

What is $\angle COD$ in terms of θ? $\angle COD =$ _____

Write the following lengths in terms of trigonometric functions of the angle θ:

$AB =$ _____ $BC =$ _____

$CD =$ _____ $OD =$ _____

Question 1: Looking at the big triangle ACD, what is $\sin\theta$ equal to? Use this to generate the double-angle formula for $\sin 2\theta$.

Question 2: Looking at triangle ACD again, what is $\cos\theta$? Use this to find the double-angle formula for $\cos 2\theta$.

Trigonometry on Other Triangles

In high school you learn the side lengths of 45°-45°-90° triangles and 30°-60°-90° triangles, and this allows you to know precisely what sine, cosine, and tangent are for these angles. For example, you know that $\sin 60° = \sqrt{3}/2$ because of the 1, 2, and $\sqrt{3}$ sides of a 30°-60°-90° triangle.

But what about other triangles? We can find exact side lengths for other triangles too if we fold up triangles that we already know!

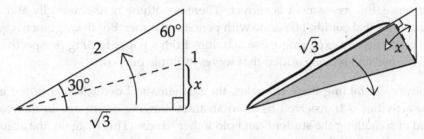

Exercise 1: Take a 30°-60°-90° triangle and fold the 30° leg up to the hypothenuse making a 15° angle. Then fold the rest of the triangle over this flap, as shown above, and unfold.

What is the length labeled x in these figures?
(Hint: Do you see any similar triangles?)

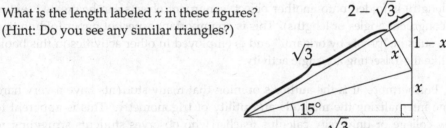

Use your answer from above to find the best exact lengths for a 15°-75°-90° triangle, where we scale the lengths to make the short side length 1. (Try to make your lengths *as simple as possible*.)

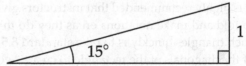

Fill in the blanks: $\sin 15° = $ _____, $\cos 15° = $ _____, $\tan 15° = $ _____.

Exercise 2: Do the same thing with a 45°-45°-90° triangle to find exact lengths of a 22.5°-67.5°-90° triangle.

SOLUTION AND PEDAGOGY

After examining and testing these handouts, instructors might have the criticism that these activities do not require origami—that they could just as easily be done by requiring students to draw careful pictures of right triangles, drawing angle bisectors or perpendicular bisectors of sides, and then exploring the trigonometry and such. Or perhaps the figures alone in these handouts could be given to the students, or the handouts themselves could be used without any folding actually done by the students! So, the critique could conclude, this isn't really an origami activity, is it?

In a sense, this assessment is correct. There is nothing mathematically about these activities that couldn't be done with pencil and paper. But the argument can also be made that by exploring these activities from a paper-folding perspective, an extra dimension is being added that serves multiple purposes:

- By physically folding these triangles, the mathematical deductions present in these activities is transferred from an abstract drawing on a page to the real world of something the student can hold in her hands. (This is, again, the argument that hands-on learning makes math more tangible.)

- Folding the triangles actually adds some logical deductions to the problems, where the student needs to observe that when one object (either an angle or a length) is folded onto another object, congruent objects can be created (either congruent angles or lengths). This is an important aspect of paper folding that results in "proofs by origami" and is employed in other activities in this book, like the Trisecting an Angle activity.

Furthermore, it is the author's opinion that many students have a very hard time internalizing the methods and utility of trigonometry. This is apparent to any college or university calculus teacher who observes students struggling to retain the trig knowledge that they encountered in high school. Any opportunity to make trig more tangible to a student can pay great dividends later on if it helps the student appreciate and remember the power of trigonometry.

Handout 1: Double-angle formulas

It is highly recommended that instructors give their students a paper right triangle to fold and make notations on as they do this activity. One easy way to generate such triangles quickly is to take standard 8.5×11 inch paper and cut it along one of the diagonals of the rectangle to create two right triangles. The angles of such a triangle are not obvious, and thus it makes for a good, general triangle with which to experiment.

The following shows an example of what students might draw on their folded triangle. First one would have to notice, as indicated on the handout, that the angle θ gets duplicated when folded to the top of the triangle. Thus $\triangle AOC$ is an isosceles triangle and BO is the perpendicular bisector of AC. Looking at right $\triangle ABO$, this immediately gives that $\cos\theta = AB/1$, and so $AB = BC = \cos\theta$.

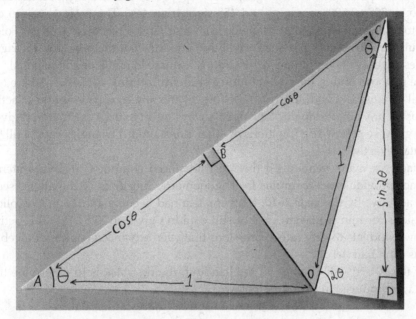

Following the angles around the triangle gives the answer to the first question on the handout, that $\angle COD = 2\theta$. That is, $\angle ACD = 90° - \theta$ from the perspective of the big triangle $\triangle ACD$, and so $\angle OCD = \angle ACD - \theta = 90° - 2\theta$. But from the perspective of $\triangle OCD$, we have that $\angle OCD = 90° - \angle COD$. Putting these together gives $\angle COD = 2\theta$.

Therefore, looking at right $\triangle OCD$, we have that $CD = \sin 2\theta$ and $OD = \cos 2\theta$. Thus, $\triangle ACD$ tells us that

$$\sin\theta = \frac{\sin 2\theta}{\cos\theta + \cos\theta} \Rightarrow \sin 2\theta = 2\sin\theta\cos\theta,$$

which is the double-angle formula for sine.

Looking at $\triangle ACD$ again,

$$\cos\theta = \frac{1 + \cos 2\theta}{\cos\theta + \cos\theta} \Rightarrow \cos 2\theta = 2\cos^2\theta - 1,$$

which is the double-angle formula for cosine. (Although the alternate form $\cos 2\theta = \cos^2\theta - \sin^2\theta$ can also be obtained by using $\sin^2\theta + \cos^2\theta = 1$.)

It should be noted that this activity is, basically, an origami version of a previously known proof of the double-angle formulas. For example, the mathematics of the proof presented here is merely simplified from that given (without paper folding) by Eli Maor in [Maor98], pp. 89–90.

Handout 2: Trig functions of nonstandard triangles

It is harder to prepare actual triangle pieces of paper for this activity, because they need to be 30°-60°-90° triangles and 45° right triangles. The easiest way to do this is to fold some equilateral triangles from a square or rectangle (see the Folding Equilateral Triangles in a Square activity) and then cut them in half; 45° right triangles are easy enough to make by cutting a square in half along a diagonal.

One purpose of this activity is to force students to think about why there are "special" triangles where the trig functions of the angles can be found exactly, whereas for any other triangles they must always resort to their calculator. It gives them a glimpse at how trig functions of other, nonstandard triangles could still be computed exactly.

In fact, for many years even the most advanced computer algebra systems could not provide exact formulas for trigonometric functions of the angles seen in this activity. By the year 2010, Mathematica had become able to give explicit expressions for things like $\sin 15°$ but still couldn't give $\sin 22.5°$. So, this activity allows students to generate expressions that even advanced computer algebra systems can't handle!

For the 30°-60°-90° triangle in the handout, the first task is to determine the length x, reproduced in the figure below.

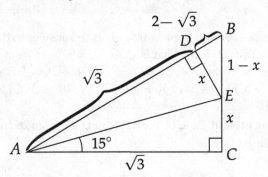

The key is that $\triangle EDB$ is similar to $\triangle ABC$ since they're both right triangles and they share the angle at B. Therefore,

$$\frac{x}{1-x} = \frac{\sqrt{3}}{2} \Rightarrow x = \frac{\sqrt{3}}{2+\sqrt{3}} = \frac{\sqrt{3}(2-\sqrt{3})}{1} = 2\sqrt{3}-3.$$

Thus our 15°-75°-90° triangle has long leg $\sqrt{3}$, short leg $2\sqrt{3}-2$, and hypothenuse $2\sqrt{6-3\sqrt{3}}$. These values are rather awkward, and normalizing the shortest leg to be 1 seems somewhat standard for this kind of thing. Doing this gives us a long leg of length $\sqrt{3}/(2\sqrt{3}-2) = 2+\sqrt{3}$ once we rationalize the denominator.

Dividing the hypothenuse by $2\sqrt{3}-3$, rationalizing the denominator, and trying to simplify the nested square roots give us a hypothenuse of length $2\sqrt{2+\sqrt{3}}$. That's not too bad, but rather amazingly, this is equal to $\sqrt{2}+\sqrt{6}$. Arriving at this is rather tricky, and it is not likely that students will be able to simplify the nested radical this far. So please feel free to accept a final solution of $2\sqrt{2+\sqrt{3}}$ for the hypothenuse! But if we let $a = 2\sqrt{2+\sqrt{3}}$, then $a^2 = 8 + 4\sqrt{3}$, and

$$8 + 4\sqrt{3} = (\sqrt{2}^2 + \sqrt{6}^2) + 2\sqrt{2}\sqrt{6} = (\sqrt{2} + \sqrt{6})(\sqrt{2} + \sqrt{6}).$$

Yes, it's not the kind of thing you'd stumble upon without experience in trying to factor quadratic surds into perfect squares. In any case, we obtain the following "canonical" lengths for a 15°-75°-90° triangle:

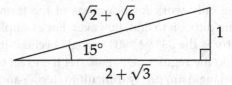

This gives us (again, after rationalizing denominators)

$$\sin 15° = \frac{\sqrt{6} - \sqrt{2}}{4}, \quad \cos 15° = \frac{\sqrt{2} + \sqrt{6}}{4}, \quad \tan 15° = 2 - \sqrt{3}.$$

The same approach handles the 45° right triangle case, and is even easier. The below figure shows that the length x gets repeated three times in the folded triangle, and so the hypothenuse gives us that $x = \sqrt{2} - 1$ without any use of similar triangles.

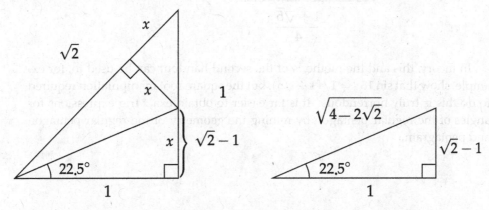

The above right figure shows the "canonical" lengths for a 22.5°-67.5°-90° triangle. Unfortunately, in this case the hypothenuse does not simplify nicely to get rid of the nested radical. As a result, the exact trig functions for 22.5° aren't especially nice:

$$\sin 22.5° = \frac{\sqrt{2 - \sqrt{2}}}{2}, \quad \cos 22.5° = \frac{\sqrt{2 + \sqrt{2}}}{2}, \quad \tan 22.5° = \sqrt{2} - 1.$$

We could normalize the triangle to make the shortest leg have length 1, but this doesn't improve things.

Further explorations

One is tempted to go beyond the double-angle formulas and try to develop origami demonstrations of, say, the angle sum formulas, like $\sin(\alpha + \beta) = \sin\alpha\cos\beta + \cos\alpha\sin\beta$. One could simply mimic an existing geometry proof, but this is equivalent to just making fold lines on a sheet of paper instead of drawing them; it could be done, and it might be fun to fold it instead of draw it, but the origami connection starts to get tenuous at some point.

In the second handout, one can explore other triangles as well, although the methods presented here only work for bisections of the triangles we know already. Other triangles are difficult to find, however. For example, another triangle that pops up in geometry is the 36°-54°-90° triangle because it is the fundamental right triangle found in the regular pentagon. Yet the exact expressions for the side lengths of such a triangle are pretty difficult to derive and don't simplify to especially nice expressions, although the golden mean is present in them. See the figure below.

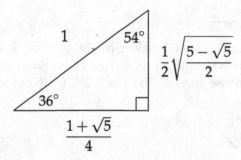

In theory, this and the methods of the second handout can be used to, for example, show that $\sin 18° = 1/(1 + \sqrt{5})$, but the square root manipulation required to do this is truly horrendous. It is far easier to obtain exact trig expressions for angles of the regular pentagon by mining the geometry of the regular pentagon and pentagram.

Activity 3
DIVIDING A LENGTH INTO EQUAL NTHS: FUJIMOTO APPROXIMATION

$$1/5 \pm E$$

For courses: calculus, number theory, discrete dynamics, modeling

Summary

Fujimoto's approximation technique for folding a strip of paper (or the side of a square) into $1/n$ths for n odd is presented. Numerous questions can then be asked, such as "Why does it work?" and "What does the sequence of left and right folds in this method tell us?" and "When do we get pinch marks at all multiples of $1/n$?"

Content

Simply teaching Fujimoto's method and seeing how it works is a great, hands-on demonstration of exponential decay, since the error in the initial guess decreases by a power of 2 at every iteration. The connection to exponential functions and analogy to things like Newton's Method make this part, alone, good for a calculus class.

To analyze Fujimoto's method in more detail, a mathematical model of the situation needs to be created. It turns out to be incredibly useful to think of the strip as the interval $[0,1]$ on the real number line and to consider the numbers we are generating in their binary decimal representation. Folding the left or right sides of the paper in half turns out to be equivalent to inserting either a 0 or a 1 at the beginning of the number's binary decimal, which establishes a specific mathematical meaning to the folds being made. Studying this can easily fit into the context of a mathematical modeling or discrete dynamical systems class.

But there is also some interesting number theory at play here. The question of knowing whether or not one will make pinch marks at every multiple of $1/n$ as Fujimoto's method is performed turns out to be equivalent to whether or not n is prime and 2 is a primitive root mod n. So this paper folding activity makes a fun applied number theory problem.

Handouts

Two handouts are provided. The questions on the second handout progress from one to the other, but they can also stand alone. (For example, Questions 4–6 could be completely skipped in a number theory class, if desired.)

(1) Introduces Fujimoto's approximation method and asks the general question of why it works.

(2) Analyzes Fujimoto's method. The first part is basic, the second is for dynamical systems, and the third is for number theory.

Time commitment

Teaching the approximation method will take only 10 minutes, but students will need another 20 at least to figure out why it works and try it for other values of n. The time needed for the second handout parts will depend largely on the class in which they're used.

How Do You Divide a Strip into Nths?

Oftentimes in origami we are asked to fold the side of a square piece of paper into an equal number of pieces. If the instructions say to fold it in half or into fourths, then it's easy to do. But if they ask for equal fifths, it's a lot harder. Here you'll learn a popular origami way of doing this, called **Fujimoto's approximation method.**

(1) Make a **guess pinch** where you think a 1/5 mark might be, say on the left side of the paper.

(2) To the right of this guess pinch is $\approx 4/5$ of the paper. Pinch this side **in half**.

(3) That last pinch is near the 3/5 mark. To the right of this is $\approx 2/5$ of the paper. Pinch this right side **in half**.

(4) Now we have a 1/5 mark on the right. To the left of this is $\approx 4/5$. Pinch this side **in half**.

(5) This gives a pinch nearby the 2/5 mark. Pinch the left side of this **in half**.

(6) This last pinch will be **very close** to the actual 1/5 mark!

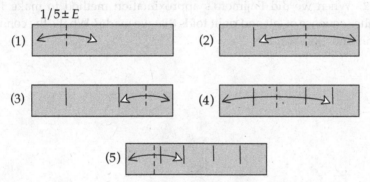

Once you do this you can repeat the above steps starting with the last pinch made, except this time **make all your creases sharp and go all the way through the paper.** You should end up with very accurate 1/5ths divisions of your paper.

Question: Why does this work?

Tip: If the strip is one unit length, then your first "guess pinch" can be thought of as being at $1/5 \pm E$ on the x-axis, where E represents the error you have. In the above picture, write in the x-position of the other pinch marks you made. What would their coordinates be?

Explain: Seeing what you did in the tip, write, in a complete sentence or two, an explanation of why Fujimoto's approximation method works.

Details of Fujimoto's Approximation Method

(1) Binary decimals?

Recall how our base 10 decimals work: We say that $1/8 = 0.125$ because

$$\frac{1}{8} = \frac{1}{10} + \frac{2}{10^2} + \frac{5}{10^3}.$$

If we were to write $1/8$ as a **base 2 decimal**, we would use powers of 2 in the denominators instead of powers of 10. So we'd get $\frac{1}{8} = \frac{0}{2} + \frac{0}{2^2} + \frac{1}{2^3}$. We write this as $1/8 = (0.001)_2$.

Question 1: What is $1/5$ written as a base 2 decimal?

Question 2: When we did Fujimoto's approximation method to make 1/5ths, what was the sequence of left and right folds that we made? What's the connection between this and Question 1?

Question 3: Take a new strip of paper and use Fujimoto to divide it into equal 1/7ths. How is this different from the way 1/5ths worked? Find the base 2 decimal for $1/7$ and check your observations made in Question 2.

(2) A discrete dynamics approach...(courtesy of Jim Tanton)

We've been assuming that our strip of paper lies on the x-axis with the left end being at 0 and the right end at 1. Let's define two functions on this interval $[0, 1]$:

$$T_0(x) = \frac{x}{2} \text{ and } T_1(x) = \frac{x+1}{2}.$$

Question 4: What do these two functions mean in terms of Fujimoto's method?

Question 5: Let $x \in [0, 1]$ be our initial guess in Fujimoto's method for approximating 1/5ths. (So x will be something like $1/5 \pm E$.) Write x as a binary decimal, $x = (0.i_1 i_2 i_3 \ldots)_2$.

What will $T_0(x)$ be? How about $T_1(x)$? Proofs?

Question 6: As we perform Fujimoto's method on our initial guess x, we can think of it as performing T_0 and T_1 over and over again to x. When approximating 1/5ths, what happens to the binary decimal of x as we do this? Use this to prove the observation that you made in Question 2.

(3) A number theory question...(courtesy of Tamara Veenstra)

In Question 3 you were asked to use Fujimoto to approximate 1/7ths, and you should have noticed that in doing so you do not make pinch marks at every multiple of 1/7, unlike when approximating 1/5ths. Indeed, only pinch marks at 1/7, 4/7, and 2/7 are made.

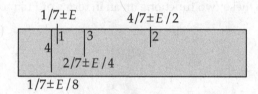

	7ths left	7ths right
	1	6
	4	3
	2	5
	1	6

We can keep track of what's going on in a table, like the one to the right. The first line shows how many 1/7ths are on the left of the first pinch and how many are on the right. The second line does the same for the second pinch, and so on. As you can see, the right side starts at 6 and comes back to 6 after only 3 lines. So it doesn't make all 1/7ths pinch marks.

Assignment: Make similar tables for 1/5ths, 1/9ths, 1/11ths, and 1/19ths:

5ths left	5ths right
1	4

9ths left	9ths right
1	8

11ths left	11ths right
1	10

19ths left	19ths right
1	18

Question 7: Think about what these tables are telling you in the number system $\mathbb{Z}_n$ (the integers mod n) under multiplication, where n is the number of divisions. Then answer the question: How can we tell whether or not Fujimoto will give us pinch marks at every multiple of $1/n$ when approximating $1/n$ths?

SOLUTION AND PEDAGOGY

Why does Fujimoto work?

The figure below shows the numbers students should assign to the pinch marks made when approximating 1/5ths.

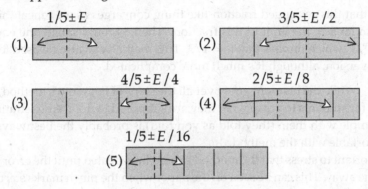

The error term in each of these pinch marks is decreasing exponentially. In other words, since $\lim_{n\to\infty} E/2^n = 0$, the error will decrease with each fold at a fast rate.

This can lead to a discussion on the nature of error when doing things in the "real world" (like origami). For example, the brilliance of Fujimoto's method is that it manages to work despite the inherent error that paper folding creates. No crease can be made with perfect, mathematical precision. No matter how hard we try, there will always be some error in our folds. So which is better, a mathematically perfect method of folding 1/5ths, or Fujimoto's approximation method? The end result will have the same amount of error; the former will have the error inherent in any fold, while the latter will have the initial error reduced to as close to zero as one's folding accuracy will allow. In fact, mathematically precise methods for folding 1/*n*ths have a habit of compounding error with each fold, whereas Fujimoto keeps reducing error with each fold. This is one reason why so many origamists use Fujimoto's method when folding odd divisions. Modeling the error in Fujimoto more accurately, to take into consideration human error as well, might make an interesting student project.

There are other ways of proving that Fujimoto's method works for certain values of *n* that do not rely on the suggestions of the handout. In fact, instructors may decide to not use the handout and simply teach their class the method via a hands-on demonstration. Then the proof can be left for students to do any way they wish, and they often do not think of the simple error-reduction argument given above.

One solution that my students developed proceeds as follows: Suppose that we were to approximate 1/3rds. When we make our initial guess (near 1/3, assuming our paper side is the interval $[0, 1]$), let the length of paper from 0 to the guess pinch be a. Then the remainder of the paper is $1 - a$, and after the second pinch this will be divided in half to give two segments of length $(1 - a)/2$. Then making the third pinch would make a segment of length $(1 - (1 - a)/2)/2$. In other words, this process is recursive; after the fourth pinch we'd have segments of length

$$1 - \frac{1 - \frac{1-a}{2}}{2}.$$

Assuming that this continued fraction-like thing converges as we repeat this process, we can let S = the limit of this fraction. Then S would satisfy the equation $S = (1 - S)/2$, which implies that $S = 1/3$. This method can also be used to solve the 1/5ths version, although it's much more complicated.

Pedagogy. While the handout goes over all the steps of Fujimoto's method, it can be difficult for students to grasp what they are supposed to do. Demonstrating the 1/5ths example with them (they fold as you fold) is probably the best way to get them comfortable with the method.

It's important to stress that the method should be repeated until the error seems to have gone away. This can be seen on the paper when the pinch marks start being made directly on top of previous pinches. That's the time to make the creases sharp and go all the way across the paper. When done, "accordion pleat" the paper into a zig-zag to demonstrate that yes, the paper is divided into equal 5ths now.

After the students work through the first handout, or if some groups or individuals finish early, challenge them to use Fujimoto to make some other divisions, such as 1/7ths or 1/9ths (or 1/3rds!). This tests whether or not they really understand the method.

Handout 2: Binary decimals

In Question 1, we have that 1/5 is less than 1/2 and 1/4, but greater than 1/8. So the first three digits in the binary decimal of 1/5 are $(.001\ldots)_2$. After we take out 1/8, we have $1/5 - 1/8 = 3/40 = .075$ left over. This is bigger than 1/16, so the fourth digit is a 1. Then we have $3/40 - 1/16 = 1/80 = .0125$ left. This is smaller than 1/32 and 1/64. But wait, $1/80 = (1/5)(1/16)$ and we got 1/80 after removing the 1/16 term from it. This means that if we factor out a 1/16 from our 1/80 remainder, we get 1/5 and we're back to where we started! So the binary decimal will repeat after the first four digits: $1/5 = (0.\overline{0011})_2$.

In Question 2, we know that in Fujimoto's method to make 1/5ths, we had to fold the right side twice and then the left side twice. So we folded the sequence Right, Right, Left, Left. Students may be tempted to let Right = 0 and Left = 1, getting the same sequence as in the binary decimal of 1/5, but there's little justification for that. It makes more logical sense to let Right = 1 and Left = 0, since we should be thinking of the strip of paper being the interval $[0, 1]$, so the right side is at 1 and the left is at 0. Then we get that the left-right folding sequence is just the repeated part of the binary decimal expansion written backwards. Actually proving this comes in the second part of this handout.

For Question 3, the difference with approximating 1/7ths is that you only get pinch marks at the 1/7, 4/7, and 2/7 spots, whereas when approximating 1/5ths we got pinch marks at *every* multiple of 1/5. This oddity gets explored in part 3 of the handout. Nonetheless, $1/7 = (0.\overline{001})_2$, and sure enough, the folding sequence for this is Right, Left, Left. This should catch any poorly-stated conjectures from Question 2.

Pedagogy. Students (and indeed, many faculty) will not know or remember how to convert a real number into a binary decimal. The example of 1/8 given in the handout provides a sufficient summary, but it is not likely to be enough for students to compute the binary decimal for 1/5. Still, instructors should let the student groups try to figure out 1/5 in binary themselves a bit before, if needed, providing hints. If everyone is lost, going over an example like $1/3 = (0.\overline{01})_2$ for the whole class might help.

It is very likely that students will conjecture incorrectly in Question 2, but that's OK. Part of the learning process in formulating conjectures from data is understanding how to check yourself and recover when you get them wrong. But students need to understand the importance of checking themselves, and Question 3 should give them the opportunity for that. Make sure that students actually revise their conjectures after Question 3, not just tear them up and let them die.

Handout 2: Discrete dynamics

This material was gleaned from ideas of Jim Tanton [Tan01]. The functions $T_0(x)$ and $T_1(x)$ are doing exactly the same things as the left and right fold operations. That is, if $x \in [0, 1]$, then $T_0(x)$ is the location of the crease pinch made when folding the left side to x. (Just divide in half!) $T_1(x)$ is the location of the pinch made when folding the right side to x. That takes care of Question 4.

In Question 5, I only mention the 1/5ths example so that the sequel in Question 6 will make more sense. The important realization is that if $x = (0.i_1 i_2 i_3 \ldots)_2$, then

$$T_0(x) = (0.0 i_1 i_2 i_3 \ldots)_2 \text{ and } T_1(x) = (0.1 i_1 i_2 i_3 \ldots)_2.$$

Proving these is pretty straightforward: Since $x = \sum_{n=1}^{\infty} i_n / 2^n$, we have

$$T_0(x) = \frac{1}{2} \sum_{n=1}^{\infty} \frac{i_n}{2^n} = \sum_{n=1}^{\infty} \frac{i_n}{2^{n+1}} = (0.0 i_1 i_2 i_3 \ldots)_2 \quad \text{and}$$

$$T_1(x) = \frac{1}{2} + \frac{1}{2} \sum_{n=1}^{\infty} \frac{i_n}{2^n} = \frac{1}{2} + \sum_{n=1}^{\infty} \frac{i_n}{2^{n+1}} = (0.1 i_1 i_2 i_3 \ldots)_2.$$

So in Question 6, we see that if the folding sequence in Fujimoto for 1/5ths is RRLL repeated, then we'd be iterating $T_0(T_0(T_1(T_1(x))))$ over and over again. Given our arbitrary initial guess $x = (i_1 i_2 i_3 \ldots)_2$, we get that $T_0(T_0(T_1(T_1(x)))) = (0.0011 i_1 i_2 i_3 \ldots)_2$. This process will continue, giving us better and better approximations to 1/5.

Thus we know why the left-right folding sequence gives us the digits in the binary decimal repeated part in reverse; it's because when composing these folding operations we're adding a digit to the beginning of the binary decimal expansion, and this reverses their order in the binary digits. It's the same thing as when students get confused by the way function composition can look "backwards" to the order in which one would state the operations verbally.

Pedagogy. The functions T_0 and T_1 brilliantly capture what the left and right folds are doing. The instructor could even ask students to invent these functions themselves; the left fold is just dividing the segment $[0, x]$ in half, and the right fold is dividing $[x, 1]$ in half.

This part of the handout is a good example of how abstract mathematical functions can mean something very real—in this case something the student is holding in her hands! So it is important that the students discover the relationship between the T_0 and T_1 functions and the Fujimoto folds themselves. It's not a difficult association to make, but students need to mentally internalize it before proceeding with the rest.

Discovering the result of Question 5 might be difficult. When in doubt, always try examples. If students get stuck, ask them, "What if $x = 1/2$? What if $x = 3/4$?" Those examples can get them thinking on the right track.

Proving the results of Question 5 requires familiarity with infinite sums and a solid understanding of binary decimals. Students in a modeling or dynamics (post-calculus) course should be able to handle this. (If not, then this will be very good practice to sharpen their basic skills!)

Question 6 involves synthesizing what is learned from Questions 4 and 5. This is an important part of the experimentation-conjecture-proof process. Make sure they write down their conclusions clearly in complete sentences.

Handout 2: Number theory

This part of the handout grew from a solution to the "What pinch marks will I get?" question by Tamara Veenstra. The correct values for the tables are below:

5ths left	5ths right
1	4
3	2
4	1
2	3
1	4

9ths left	9ths right
1	8
5	4
7	2
8	1
4	5
2	7
1	8

11ths left	11ths right
1	10
6	5
3	8
7	4
9	2
10	1
5	6
8	3
4	7
2	9
1	10

(19ths is left for you.) If we read these tables backwards, from the bottom up, it's hard to miss the appearance of taking 2 to greater and greater powers in $\mathbb{Z}_n$, where n is the number of divisions we're making in Fujimoto. This makes sense too—once you make a fold in Fujimoto that results in a number of divisions equal to a power of 2 on one side of the pinch, then you'll use that side for the rest until you get to either $1/n$ or $(n-1)/n$. So, if the consecutive powers of 2 in $\mathbb{Z}_n$ generates all the numbers from 1 to $n-1$, then we'll get pinch marks at every multiple of $1/n$. In other words, the condition we're looking for is if 2 generates the set $\mathbb{Z}_n \setminus \{0\}$ under multiplication. The most concise way of saying this, which any number theory student should try for, is that 2 is a primitive root of $\mathbb{Z}_n$.

Depending on the amount of class time that you devote to this, or on the level of your students, an informal explanation like that given above may be appropriate. But it can be made much more rigorous, as follows.

Let $1/n = (0.\overline{i_1 i_2 \ldots i_k})_2$. This means that

$$\frac{1}{n} = \sum_{j=0}^{\infty} \left(\frac{i_1}{2^{jk+1}} + \frac{i_2}{2^{jk+2}} + \cdots + \frac{i_k}{2^{jk+k}} \right) = \sum_{j=0}^{\infty} \frac{2^{k-1}i_1 + 2^{k-2}i_2 + \cdots + 2^0 i_k}{2^{jk+k}}.$$

Let $a = 2^{k-1}i_1 + 2^{k-2}i_2 + \cdots + 2^0 i_k$, the numerator term of the last summation. Then notice that $a = (i_1 i_2 \ldots i_k)_2$, i.e., the number that we get from the repeating part of $1/n$ considered as an integer base 2. Also notice that a is not dependent on j. Thus,

$$\frac{1}{n} = a \sum_{j=0}^{\infty} \frac{1}{2^{jk+k}} = \frac{a}{2^k} \sum_{j=0}^{\infty} \frac{1}{2^{jk}}$$

$$= \frac{a}{2^k} \frac{1}{1 - 1/2^k} = \frac{a}{2^k} \frac{2^k}{2^k - 1} = \frac{a}{2^k - 1}.$$

We've written $1/n$ as a fraction with one less than a power of two in the denominator. This means that

$$an = 2^k - 1 \text{ or, in other words, } 2^k \equiv 1 \pmod{n}.$$

So 2 is in the group of units $U(\mathbb{Z}_n)$. Also, suppose that k is not the smallest positive integer satisfying $2^k \equiv 1 \pmod{n}$. Then we could write $1/n = b/(2^m - 1)$ for some positive integers b and $m < k$. But then we could do the above calculations backwards and get that $1/n$ has a different binary decimal expansion with a shorter repeating part. Assuming k already gave us the shortest repeating decimal, this won't happen. So k must be the smallest positive integer with $2^k \equiv 1 \pmod{n}$.

We can now interpret all this as follows: Approximating $1/n = (0.\overline{i_1 i_2 \ldots i_k})_2$ using Fujimoto will generate pinch marks at all multiples of $1/n$ if and only if $k = n - 1$, which will be true if and only if the powers of 2 generate all of $\mathbb{Z}_n \setminus \{0\}$. In other words, n must be prime and 2 must be a primitive root modulo n.

Pedagogy. Depending on when this is done in a number theory class, students will have mixed success finding the best way to state their findings. But the pattern that the tables reveal should be clear. This provides a good test for how competent the students are becoming at number-theoretic pattern-matching. Non-number theory courses can have fun with this activity as well, but the students will need to be familiar with $\mathbb{Z}_n$. (And they're not likely to state anything about primitive roots.)

One easy mistaken conjecture for students to make is that we'll get all the pinch marks if n is prime. Such thinking usually goes like this: if powers of 2 do not generate all of $\mathbb{Z}_n \setminus \{0\}$, then they'll generate a subgroup, and the only time when $\mathbb{Z}_n$ won't have any subgroups is when n is prime. This is flawed, of course, but don't be surprised if you see students thinking in this way.

Further studies

Shuzo Fujimoto describes his approximation method in the extremely rare Japanese book [Fuj82]. He also used this technique as a way to approximate odd angle divisions. For a different look at this type of approximation method and other tie-ins to number theory, see the work of Hilton and Pedersen, such as [Hil97].

Activity 4
DIVIDING A LENGTH INTO EQUAL *N*THS EXACTLY

(1) 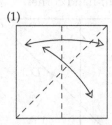 (2) 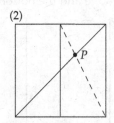 (3) 1/3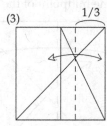

For courses: geometry, precalculus

Summary

Students are asked to come up with ways to, say, fold the side of a square piece of paper into perfect 3rds or 5ths or some other odd division. The aim here is to develop **exact** methods, not approximations.

After students have tried this for a while, or perhaps in a later class, give them the handout. This shows an origami routine that the students will discover produces a landmark for folding perfect 1/3rds. The students are then asked to generalize this method.

Content

This activity is mostly geometry, although it's a problem that can be solved using both synthetic and analytic methods. In fact, if the problem is solved analytically, nothing more than finding equations of lines and their point of intersection is used, making this a nice hands-on activity for a precalculus class.

Handouts

There are two handouts that take two different approaches to the same task: folding a square piece of paper into perfect thirds. The first one shows students the folding method and challenges them to discover what it is doing. The second one explains what the method is doing and challenges them to prove it.

Both of these handouts can be motivated by asking students beforehand to try coming up with their own methods of folding thirds exactly.

Time commitment

Plan on reserving at least 30 minutes of class time for this activity, which includes folding time, student work time, and discussion afterward.

What's This Fold Doing?

Below are some origami instructions. Take a square and make creases by folding it in half vertically and folding one diagonal, as shown. Then make a crease that connects the midpoint of the top edge and the bottom right-hand corner.

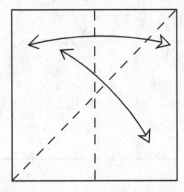 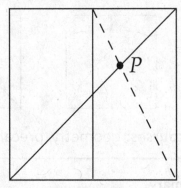

Question 1: Find the coordinates of the point P, where the diagonal creases meet. (Assume that the lower left corner is the origin and that the square has side length 1.)

Question 2: Why is this interesting? What could this be used for?

Question 3: How could you generalize this method, say, to make perfect 5ths or $n$ths (for n odd)?

Folding Perfect Thirds

It is easy to fold the side of a square into halves, or fourths, or eighths, etc. But folding odd divisions, like thirds, exactly is more difficult. The below procedure is one was to fold thirds.

(1) 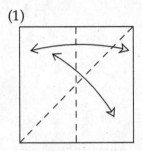 (2) 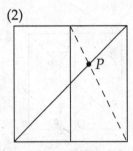 (3) 1/3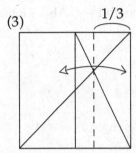

Question 1: Prove that this method actually works.

Question 2: How could you generalize this method, say, to make perfect 5ths or *n*ths (for *n* odd)?

SOLUTION AND PEDAGOGY

Since the two handouts are similar, we'll focus on solutions for the first one.

Question 1: Synthetic approach

Assume that the square has side length one and consider the labeling in the figure below. Denote the coordinates of P with (x, x). Then AE has length x, so EB has length $1 - x$. Also, EP has length x.

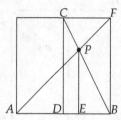

Note that $\triangle BDC$ and $\triangle BEP$ are similar. Thus $|CD|/|PE| = |BD|/|BE|$, which becomes

$$\frac{1}{x} = \frac{1/2}{1 - x} \Rightarrow 2 - 2x = x \Rightarrow x = \frac{2}{3}.$$

This could also be proven by using the similar triangles $\triangle ABP$ and $\triangle CPF$.

Question 1: Analytic approach

Assume that the square sits in the xy-plane, with A at the origin and B at $(1, 0)$. Then P lies on the intersection of two lines: $y = x$ and $y - 1 = -2(x - 1/2)$. Combining these to find their intersection gives $x - 1 = -2x + 1$, or $3x = 2$, or $x = 2/3$.

Obviously, the answer to Question 2 is that this can be used to fold the square into thirds exactly. Try it!

Question 3

The picture below shows how to generalize this method to fold the side of a square into n equal divisions, where n is odd. Instead of using the $1/2$ vertical line, make a vertical line at $x = (n - 2)/(n - 1)$ (or $1/(n - 1)$ away from the right side).

$1/(n{-}1)$

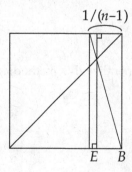

Finding this line should not be too hard, since $n - 1$ is an even number (since n is odd). (If $n - 1$ equals something like 6, then you'd have to find a $1/3$ point first and then fold this in half to get a $1/6$ mark. So in a sense this method is recursive.)

The same approaches to Question 1 will give that the point at which the two diagonal creases in this general case meet is $((n - 1)/n, (n - 1)/n)$, which can then be used to fold the paper into equal $n$ths.

Pedagogy

As mentioned previously, students appreciate learning methods of folding perfect thirds a lot more when they've spent some time themselves trying to develop them. There are many other methods for doing this kind of thing (some of which will be described at the end of this section), and if students come up with methods of their own then they should be studied and proven. In fact, if someone comes up with the method provided in the handout, then that's the best context in which to investigate proofs and generalizations. Thus, if the students' own explorations go well, there may be no need for the handout.

The first handout may seem more advanced, but I've been surprised at how able some students are at figuring out what the method is doing. Nonetheless, the first handout does set students up for an analytic proof, since finding the coordinates of P is most easily done by finding the equations of the crease lines.

The second handout places more emphasis on developing proof-building skills. Most students come up with the similar triangles proof, but the analytic approach can be a very useful one in a variety of geometry problems and uses nothing more than basic precalculus material. In a geometry course students are often delighted to learn that they can solve some problems using such simple techniques. So if all groups develop synthetic geometry proofs, make sure to drop some hints to students who finish early about thinking of the paper as being in the xy-plane, so that equations of the lines can be found. Usually this is all that needs to be told for students to run with this and develop the analytical proof described above. (And note that the second handout gives no hints about an analytic proof, unlike the first handout.)

The general method is also easy for students to figure out, if they first try a simple case. Students who are stumped on how to generalize should be encouraged to try an example, like folding $1/5$ths. To make $1/5$ths with this method requires only using a vertical line at the $x = 3/4$ position instead of the $1/2$ position. This is pretty straightforward for students to figure out and can lead to the complete generalization.

Other methods

As mentioned previously, there are many other methods for folding $1/3$rds, $1/5$ths, or general $1/n$ths. A few will be shown here without proof.

Below is shown a way to achieve $1/3$rds that follows naturally from one of the methods for folding a $30°$-$60°$-$90°$ triangle (as seen in Activity 1). This does not generalize to other $1/n$ths, however.

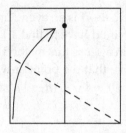 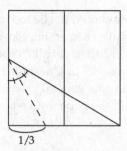

1/3

A different general method, shown below, was invented by Haobin Yu, a student at the 2000 Hampshire College Summer Studies in Mathematics. It uses the premise, again, that divisions of $1/2n$ should be possible, and from this we get an odd division $1/(2n+1)$. It can be proven using similar triangles or the analytic method used previously.

$1/(2n)$

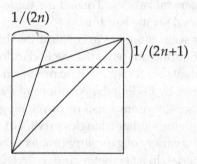

$1/(2n+1)$

More methods for folding exact divisions can be found via web searches, in particular see [Hat05] and [Lang04-1]. Any of these methods could be assigned for homework exercises or extra projects.

Activity 5
ORIGAMI HELIX

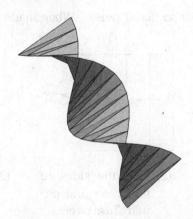

For courses: trigonometry, precalculus, calculus, geometry

Summary

An origami helix is folded that leads to an exercise where we ask, "How much can this folding technique cause a square piece of paper to be rotated in the limit?"

Content

The model here is a clever way to make paper twist using a simple pleat technique. The end result looks like a helix and resembles wind spinners made of wood that are used as outdoor home decorations. The exercise in this activity is a natural extension of the folded model and uses the concept of radian angle measure, geometric transformations, trigonometric functions, and limits. The most simple solution to the problem, however, avoids direct use of limits but does require a firm understanding of radian angle measure. As such, this activity is appropriate for classes ranging from trigonometry through calculus.

Handout

The first page of the handout shows the instructions for the model, and the second page asks the question of how much does the square piece of paper twist in the "limit" of this model.

Time commitment

Folding the version of this model as shown in the instructions takes a class about 15 minutes. Allow another 15 minutes for answering the question. However, if students distract themselves by obtaining longer strips of paper to make a larger helix, things will take longer.

Folding a Helix

This model pleats the paper so that it twists. When made from a long strip the result is a helix.

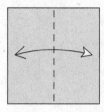 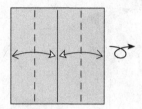 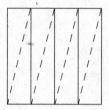

(1) Fold and unfold in half, from side to side.

(2) Fold the sides to the center and unfold. Turn over.

(3) Now carefully fold diagonal creases in each rectangle.

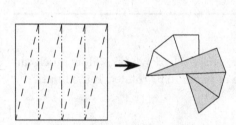 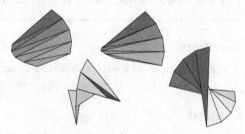

(4) Now fold all the creases at the same time. The result will be a square that has been twisted.

If you let the model be 3D, it makes an interesting shape!

Folding this model from a strip of paper makes a twisted helix shape, as shown on the left. (You need to make a lot more divisions along the strip for this to work.)

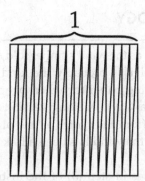

Question: If we made more divisions in steps (1)–(2) in the above instructions, we would get more of a twist from our square. Below is a row of examples made with only 3 divisions as in steps (1)–(2), with 6 divisions, with 8 divisions, and with 13 divisions. In each the angle α is slowly getting smaller!

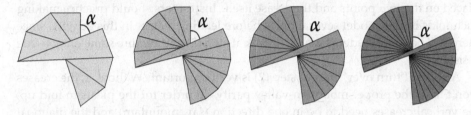

So the question is, what happens to this angle α as we make more and more divisions?

Or, putting it another way, how much does the square twist as we make more and more divisions? Will it keep twisting more and more, or does it approach a limit?

SOLUTION AND PEDAGOGY

This origami helix, also known as the wind spinner fold, is considered a traditional model. This simple pleat design is probably the kind of thing that was folded when people first started folding paper (or when people first had good enough quality paper to fold in such a way). Making versions of this design from long rectangles and hanging them from the center of one of the short sides to make a helix that spins attractively in the wind has been a folk craft activity for at least a century, and probably longer.

Making the diagonal creases in step (3) of the instructions can be tricky for students (and instructors too, perhaps). This is because making such creases is, literally, performing the origami move where we have two points on the paper and we want to make a straight crease that connects them. There is no landmark as to where the rest of the paper has to go when doing this, and so the attention is placed on the two points and the crease itself. Instructors should practice making such folds in this model several times before leading a class in this activity, since there will be some students who will balk at step (3) and require some one-on-one assistance.

Also, the "turn over" part of step (2) is very important. Without it, the creases won't have the proper mountain-valley parity. In order for the pleats to fold up, the vertical creases need to be in one direction (say, mountain) and the diagonal creases need to be in the other (valley).

One danger in this activity is for students to become distracted by wanting to fold longer helixes and never get to the second page of the handout, where the math question resides. As an avid paper folder, the author has a hard time dissuading such enthusiasm from his students.

Also, some students might not believe the math question and claim that the angles α are all the same in the pictures, or in their folded models. If, however, a model with very few divisions (like 3 or 4) is placed on top of a model with lots of divisions, you can see that the angle is *very slowly* getting smaller, as the below figure tries to illustrate. (Note that one's models need to be folded quite accurately to confidently see this difference in practice.)

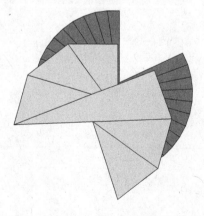

The solution to the question has a very simple solution as well as more complex approaches that can be taken. Only very bright students will see the easy solution quickly, so instructors can mostly expect their students to take a more methodical tact. Therefore, we will present the more complex solution first.

First solution (harder)

In order to quantify how much the paper rotates, the crease pattern needs to be examined carefully. Specifically, each pleat, where we have a vertical mountain crease and then a diagonal valley crease next to it, rotates a section of the paper, and with multiple pleats this rotation adds to itself. A pair of creases that makes such a pleat is shown below.

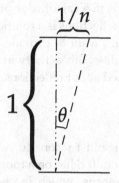

Now, each crease is being folded flat, and therefore we can think of each crease line as *reflecting* part of the paper. Thus, the combination of these two creases is that they perform two reflections on the paper. The product of two reflections that meet at an angle θ, as ours do above, is equal to a rotation by an angle of 2θ. Therefore, the angle θ that is determined by these pleats will determine our total rotation of the paper.

Since the two creases shown above form a right triangle, if we've made n divisions of our square in steps (1)–(2) of the instructions, then we have

$$\tan \theta = \frac{1/n}{1} \Rightarrow \theta = \arctan(1/n).$$

We have n of these pleats in the square, and each rotates the paper by 2θ. Thus our total amount of rotation is

$$2n \arctan(1/n),$$

and we just need to take the limit of this as n goes to infinity. Students with a calculus background should be able to handle this limit. We can first substitute $x = 1/n$ and then apply L'Hospital's Rule:

$$\lim_{n \to \infty} 2n \arctan(1/n) = \lim_{x \to 0} \frac{2 \arctan x}{x} = \lim_{x \to 0} \frac{2}{1 + x^2} = 2.$$

This is very surprising! The answer is that in the limit, the square piece of paper will make a twist of 2 radians.

How often does one see a problem in a precalculus or calculus class for which the answer is a whole number of radians? Usually problems in such classes are a multiple of π radians. This helix-folding problem is novel in that the answer turns out to be exactly a whole number of radians.

Of course, students will probably want to convert this to degrees (getting about $114.592°$). However, students will likely be a bit confused by the answer of 2: 2 what? Unless students remember that in calculus we need to think of functions like arctan as returning values in radian measure, they might think that the units of the final answer are degrees! This can provide a good learning moment, or review, of the central role that radian measure plays in analysis.

Admittedly, however, students in a calculus or precalculus class will likely be unaware that the product of two reflections is a rotation, let alone that this rotation would be by 2θ. Instructors might want to provide this information as a hint to such classes. But in a geometry class, this activity makes for a nice application of what students should have learned about reflections.

Second solution (easier)

The easier solution is due to origamist Jason Ku, who came up with it spontaneously over a dinner conversation. It relies on a thorough understanding of what radian angle measure actually means, which the below-left illustration shows. That is, 1 radian is the angle that cuts out an arc of length 1 from the center of a circle of radius 1.

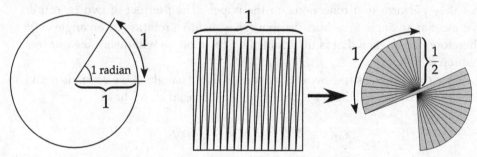

Keeping this definition of radian angle measure in mind, consider our square piece of paper to have side length 1. When we fold this square into the helix with some number of divisions, the side of the square approximates an arc of a circle of length 1. But as the above-right figure illustrates, this arc of length 1 in the flattened helix is from a circle with radius equal to 1/2. Of course, for any given helix model, these measurements are approximations, since our "arc of length 1" is not really a perfectly circular arc, and the actual radius might not be exactly 1/2 if we measured it. However, these are clearly the lengths that we would be approaching in the limit case.

In any case, the amount of rotation that the square of paper undergoes is exactly the angle that subtends our arc of length 1. This is not an angle of 1 radian, however, since the circle here is radius 1/2. On such a circle, 1 radian would cut out an arc of length 1/2. So an arc of length 1 would be cut from an angle of 2 radians, which is exactly the answer that we obtained from the previous solution.

This solution would clearly benefit a class learning radian angle measure. Showing students figures like the one above might help them discover the fundamental connection between radian angle measure and the origami helix that, after all, explains why we get an answer that is a whole number of radians.

Further explorations

The helix is a very interesting shape, and students should be encouraged to make longer helixes. If, in doing so, one brings the two ends of the paper together and overlaps them, then an hourglass shape can be formed. The below-left picture is one possibility.

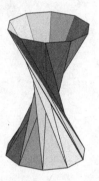
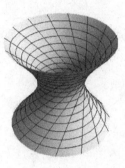

This hourglass shape seems to resemble a mathematical surface (shown above-right) called a *hyperboloid of one sheet*. A hyperboloid of one sheet is a surface given by the equation $x^2 + y^2 - z^2 = 1$. It is an example of a *ruled surface*, which means that it can be made by sweeping a straight line through 3D space. The diagonal lines shown in the above right picture are depicting how this straight line would travel, and the diagonal creases in our helix-hourglass seem to be doing the same thing. With more and more creases (i.e., more and more divisions made in our paper), these diagonal creases would approximate a hyperboloid of one sheet surface more and more.

Activity 6
FOLDING A PARABOLA

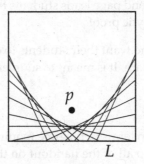

For courses: geometry, precalculus, calculus, abstract algebra, modeling

Summary

Students are led though an exercise of applying a basic origami move (fold a point to a line, which is a required move in Activity 1) over and over again to produce crease lines that seem to be tangents to a parabola. Students are asked to prove that this is, indeed, a parabola.

Follow-up activities: Can we fold an ellipse or hyperbola in similar ways? What does this tell us about the field extension of the rationals that origami constructions generates?

Content

While this is clearly a geometric construction exercise, in the sense that it's an opener to a bigger question of what geometric constructions are possible via origami as opposed to, say, straightedge and compass, there's a lot more going on here as well. Basic facts about parabolas are reinforced, and the whole proof can be done using only logic and precalculus techniques. On the other hand, providing a rigorous proof does involve creating a detailed model of the folding process in this activity, making this a good example of geometric modeling. Furthermore, a more elegant proof can be made using envelopes of curves, a topic sometimes encountered in differential or algebraic geometry courses. Thus there is a wide range of courses in which this activity can be useful.

This activity also offers a chance to illustrate, by example, the connection between visual geometry and solving algebraic equations. That is, one of the punchlines of this activity is being able to say, "Doing this origami fold is equivalent to solving a quadratic equation." This kind of connection between geometry and algebra is an important concept in higher mathematics.

Handouts

There are two handouts for this activity.

(1) The first page leads students through the parabola-folding activity and asks them to prove it. The second page leads students through modeling this kind of folding to a more analytic proof.

(2) This is for instructors who want their students to explore this exercise using dynamic geometry software. It is meant to supplement the first handout, not replace it.

Time commitment

This activity is rather flexible in terms of the time required because instructors can choose either to let students do all of the handout on their own or to incorporate parts of it into a lecture. The first page of the handout will take 10–15 minutes, and for students to work through the second page will require an additional 20–30 minutes.

The dynamic geometry software handout takes 10–15 minutes.

Exploring a Basic Origami Move

Origami books display many different folding moves that can be made with paper. One common move, especially in geometric folding, is the following:

> Given two points p_1 and p_2 and a line L, fold p_1 onto L so that the resulting crease line passes through p_2.

Let's explore this basic origami operation by seeing exactly what is happening when we fold a point to a line.

Activity: Take a sheet of regular writing paper, and let one side of it be the line L. Choose a point p somewhere on the paper, perhaps like below. Your task is to fold p onto L over and over again.

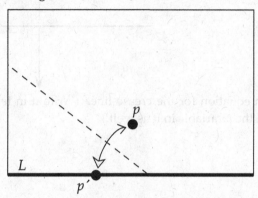

It is easier, actually, to fold L to p, by bending the paper until L touches p and then flattening the crease. Do this many times—as many as you can stand!—choosing different points p' where p lands on L.

Question 1: Describe, as clearly as you can, exactly what you see happening. What are the crease lines forming? How does your choice of the point p and the line L fit into this? Prove it.

Now we'll try to find the equation for the curve you discovered.

First, let's define where things lie on the xy-plane. Let the point $p = (0, 1)$ and let L be the line $y = -1$. Now suppose that we fold p to a point $p' = (t, -1)$ on the line L, where t can be any number.

Question 2: What is the relationship between the line segment $\overline{pp'}$ and the crease line? What is the slope of the crease line?

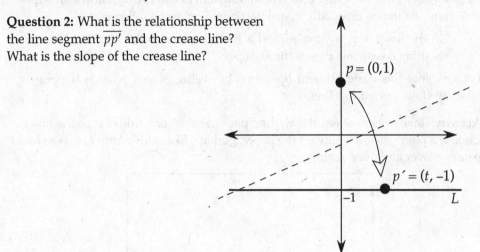

Question 3: Find an equation for the crease line. (Write it in terms of x and y, although it will have the t variable in it as well.)

Question 4: Your answer to Question 3 should give you a **parameterized family** of lines. That is, for each value of t that you plug in, you'll get a different crease line. For a fixed value of t, find the point on the crease line that is **tangent** to your curve from Question 1.

Question 5: Now find the equation for the curve from Question 1.

Origami with Geometry Software

In this activity we'll use geometry software, like Geogebra or Geometer's Sketchpad, to explore a basic origami move:

> Given two points p_1 and p_2 and a line L, fold p_1 onto L so that the resulting crease line passes through p_2.

We'll explore this basic origami operation by modeling on the software what happens when we fold a point to a line. We'll make use of a key observation:

When we fold a point p to a point p', the crease line we make will be the _____ **of the line segment** _____.

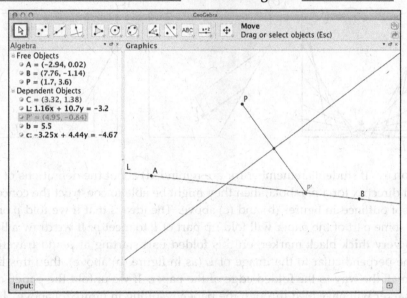

Instructions: Open a new worksheet in your software (above is shown Geogebra).

(1) Draw a line AB and label it L.

(2) Make a point not on L, call it p.

(3) Make a point on L, call it p'.

Then, with the key observation above, use the software's tools to draw the crease line made when folding p to p'.

Once you've done this, select the crease line and turn on **Trace** of the line (in Geogebra, CTRL-click or right-click on the line to do this). Then you can move p' back and forth across L and make many different crease lines. In this way you can make software do the "folding" for you! (Plus, it looks cool.)

Follow-up: What happens if we use a circle instead of the line L?

SOLUTION AND PEDAGOGY

Handout 1

The history of this folding exercise goes pretty far back. The oldest reference to it that I have found is T. Sundara Row's book *Geometric Exercises in Paper Folding*, first printed in India in 1893 [Row66]. Numerous references to "folding a parabola" in this way can be found in the math teaching literature since then. (See, for example, [Lot1907], [Rupp24], and [Smi03].) However, it never fails to surprise students (and faculty) that the outline of a parabola seems to form when repeatedly folding a point p onto a line L. Actually, the crease lines seem to be tangent to the parabola with focus p and directrix L. (See figure (a) below.)

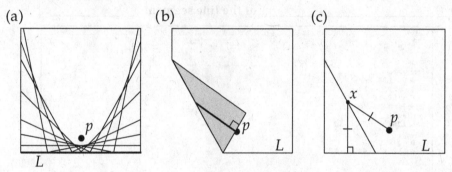

Question 1. If students remember (or are reminded) about the definitions of focus and directrix for a parabola, then they might be able to construct the conceptual proof outlined in figures (b) and (c) above. The idea is that if we fold p onto L, then some part of the paper will fold up part of L to meet p. If we draw a line, using a **very thick** black marker, on this folded part starting at p and traveling on a line perpendicular to the image of L (as in figure (b) above), then this line will run all the way to the folded edge of the paper. If we unfold the paper, our thick marker will have bled through the paper, resulting in figure (c) above. This demonstrates how the point x, where our marker line hit the crease line, is equidistant from the point p and the line L. (Recall that the distance from a point to a line is the perpendicular distance.) Furthermore, this point x is the *only* point on the crease line that will have this property. (If we try this with some other point on the line, when we refold the crease we see that there's no way the distances can be equal.) Since one definition of a parabola is the set (i.e., locus) of all points that are equidistant from a point (the focus) and a line (the directrix), we have just proven that the crease line will be tangent to the parabola with focus p and directrix L. Since our choice for where to fold p to L was arbitrary, this will hold for all our crease lines.

Question 2. When we fold p onto p', the crease line formed will be the perpendicular bisector of the line segment $\overline{pp'}$. This is fairly obvious, but some students may want to prove it rigorously. It's really the same concept at play as the elementary geometry fact that the set of all points that are equidistant from point p and from point p' is the perpendicular bisector of $\overline{pp'}$. Since the fold places one side of the paper onto the other, and p goes to p', it's clear that all points on the crease line will be equidistant from p and p'.

The slope of the segment $\overline{pp'}$ is $-2/t$, so the slope of our crease line will be $t/2$.

Question 3. The midpoint of $\overline{pp'}$ is $(t/2, 0)$, and this point will be on our crease line. Thus the equation of our crease line when folding $p = (0, 1)$ onto $(t, -1)$ will be

$$y = \frac{t}{2}\left(x - \frac{t}{2}\right) \Rightarrow y = \frac{t}{2}x - \frac{t^2}{4}.$$

Question 4. This question "gives away" the fact that our crease lines are actually tangents to a curve, but students should have figured this out in Question 1. They should also have either conjectured or proven that this curve is a parabola by now. With this piece of information, we can see that if we draw a vertical line from $p' = (t, -1)$ to the crease line at point, say, q, then folding along the crease line will show that the segments $\overline{qp'}$ and $\overline{qp}$ have the same length. Thus the crease line is tangent at point q to the parabola with focus p and directrix $y = -1$. Since q is on the crease line and we have the equation for this crease line, we see that the coordinates for q are $(t, t^2/4)$.

Question 5. There is more than one way to do this part, but Question 4 should lead students to the easiest solution. Notice that the point of tangency $(t, t^2/4)$ is actually a parameterization of the parabola $y = x^2/4$, and students can see this by merely letting $x = t$ (which makes sense since t is the x-coordinate of the point of tangency).

Notice that this solution does assume that our curve is a parabola. However, the work in Question 4 can be extended slightly to provide a proof that the curve is, indeed, a parabola. There are other ways to derive the equation of the curve, however.

Quadratic formula way. Observant students might have noticed that this basic folding operation (as stated on the first page of the handout) does not always work. That is, for a fixed p_1 and fixed L, there are choices of p_2 that would make the operation impossible to do. This can be seen in our folding exercise in that if we chose p_2 to be in the convex hull of the parabola, then no crease formed by folding p onto L could possibly go through p_2.

This can be used to our advantage: If we take our parameterized family of crease lines and *solve for* t, then we would get a formula that would tell us what value of t we should use to make sure that the crease passes through (x, y). Watch what happens when we do this:

$$\frac{t^2}{4} - \frac{t}{2}x + y = 0 \Rightarrow t = \frac{x/2 \pm \sqrt{(x/2)^2 - y}}{1/2}.$$

This gives a real number for t only when $x^2/4 - y \geq 0$. So the inequality $y \leq x^2/4$ represents all points in the plane that can be hit by a crease line, and the region given by $y > x^2/4$ contains all points that won't be hit by a crease line. The boundary of these two regions is our curve, the parabola $y = x^2/4$.

Another "assume it's a parabola" way. This proof is given by Smith in [Smi03]. Let (x, y) be the point on the crease line that is tangent to the parabola. Then notice that we must have $x = t$, where $p' = (t, -1)$ is the point on L that we fold p to. This follows from the definition of the parabola, although the illustration below makes it more clear.

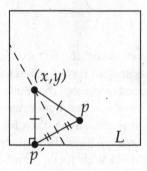

We know that the slope of our crease line is $t/2$, but this should equal the slope between points (x, y) and $(t/2, 0)$ (the midpoint of $\overline{pp'}$). So,

$$\frac{t}{2} = \frac{y - 0}{x - t/2} \Rightarrow \frac{x}{2} = \frac{y}{x/2} \Rightarrow y = \frac{x^2}{4}.$$

Calculus way. Following the previous solution, we make the observation that the slope at the point (x, y) on the parabola is $t/2$. But $x = t$, so the slope of the parabola at (x, y) is $x/2$. That is,

$$\frac{dy}{dx} = \frac{x}{2} \Rightarrow y = \frac{x^2}{4} + C,$$

which we get by taking the antiderivative. Since we know that our parabola passes through the point $(0, 0)$, $C = 0$ and our parabola has equation $y = x^2/4$.

Envelope way. (This is presented in [Huz89].) A more advanced class could look at the parameterized family of lines and realize that all we need to do is take the envelope of this family. (See [Cox05].) Specifically, if $F(x, y, t) = 0$ is a parameterized family of curves, then the *envelope* of this family (a curve that is tangential to members of the family) is given by solving the set of equations

$$F(x, y, t) = 0 \quad \text{and} \quad \frac{\partial}{\partial t} F(x, y, t) = 0.$$

In our case, we have $(\partial/\partial t)F(x, y, t) = x/2 - t/2 = 0$, or $x = t$. Plugging this into the line equation, we get $y = x^2/2 - x^2/4$, or $y = x^2/4$, which is our parabola.

Pedagogy

As students do the folding exercise, make sure they make enough creases, for otherwise they won't be able to see the parabola. It can also be useful to do the dynamic geometry software activity (see below) in between Questions 1 and 2, since simulating this on, say, Geogebra (go to http://www.geogebra.org to download this free, excellent software) helps reinforce the geometric relationships between the point p, the line L, the crease line, and the parabola. But the best way to help students develop the conceptual proof for Question 1 is for them to play with the folded paper. However, instructors will most definitely have to go over the focus-directrix definition of a parabola with the class. Most students will have forgotten this, and it's unclear how much this is emphasized in the high-school math curriculum nowadays.

Questions 2 and 3 shouldn't give students too much trouble, and they reinforce the concepts of slope, perpendicular lines, and the point-slope formula for a line. However, the answer to Question 3 is a family of lines parameterized by the variable t. Students often have trouble wrapping their mind around parameterizations, and the presence of the t variable will challenge their understanding of the finer points of this problem. It is important that they understand that the crease line is an equation in terms of x and y, like normal, but parameterized by t. The final equation for the parabola is in terms of x and y and should be more familiar-looking to students.

Questions 4 and 5 are conceptually tough. The idea is for students to find the point of tangency on the crease line, which then makes finding the equation of the parabola a snap. If this doesn't work, however, instructors should keep the other proofs in mind and offer appropriate tips as students wrestle with Question 5.

The proof that uses the discriminant in the quadratic formula to determine which region of the plane cannot contain any crease lines is the most illustrative of what's really going on in this origami operation, so I recommend going over this approach with students. (Students may also be intrigued to see such an unusual application of the quadratic formula!) This could easily become a homework or project assignment, as it might be too much to expect students to digest the folding activity and develop a proof for Questions 4 and 5 in only one class period.

This activity also drives home the point that conic sections are the locus of points satisfying a certain condition. Seeing such a hands-on illustration of this can be especially helpful for pre-service mathematics teachers, since parabolas, ellipses, and hyperbolas are still a part of the high-school algebra curriculum.

The higher-level punchline concept for this activity is that, "Origami can solve quadratic (second degree) equations." Students who fully grasp and understand this statement will leave this exercise with a lot more mathematical maturity than they brought into it. The idea that what we do in one field of math (like the geometry of paper folding) can be identical to something completely different-looking in another field of math (solving quadratic equations) is a theme that runs throughout all of mathematics. Plus, this situation is very analogous to the classic problem of trisecting an angle using straightedge and compass, where we learn that such a general construction is impossible because the tools of straightedge and compass can only solve quadratic equations, and angle trisection requires solving cubic equations. The origami parabola activity is actually a first step in seeing that origami not only can construct anything that straightedge and compass can but also can do more. This topic is pursued more in the two activities following this one.

However, proving that our activity produced a parabola does not prove that all quadratic equations can be solved via origami. It is good evidence, however, and it gives us everything we need to make a more general argument. I present an outline of this here because using the activity to launch a discussion on these topics, especially in an abstract algebra class where geometric constructions are covered, will often result in a student asking, "But how do you know that *any* quadratic equation or parabola can be solved by origami?" The following makes a convincing argument and was gleaned from Alperin's more in-depth paper on the topic [Alp00].

Proving that origami can solve general quadratic equations. The quadratic formula tells us that if you know how to perform the operations of addition, subtraction, multiplication, division, and square roots, then you know how to find the roots of any quadratic equation. Algebraically, this would be proving that the set of all points in the plane that can be constructed via origami contains the smallest subfield that is closed under square roots.[1]

Assuming that our paper is infinite (just to make our life easier) and that we start off with, say, line segments of unit length on the x- and y-axes, it is straightforward to see that addition, subtraction, multiplication, and division by rationals can be handled by origami. Adding and subtracting lengths of line segments is easy to do via folding. Division is a bit trickier, but the Dividing a Length into Equal $n$ths Exactly activity in this book proves that this kind of thing can be done. Multiplication by rationals is then just an extension of addition and division. Taking square roots is the only operation that may require the power of the parabola-inducing origami operation that we've been studying.

Suppose that r is a number (or rather, length of a line segment) that we have already constructed by folding, and we want to construct $\sqrt{r}$ somehow. We will use the construction setup described above, where we let $p_1 = (0, 1)$ be our focus and L the line $y = -1$ be our directrix. We will let our second point be $p_2 = (0, -r/4)$ and fold a crease that places p_1 onto L (at the point $p_1' = (t, -1)$) while making the crease go through p_2. We already know that the equation of our crease line is $y = (t/2)x - t^2/4$, and this line has to go through the point $(0, -r/4)$. Plugging this point into the line, we get $-r/4 = -t^2/4$, or $t = \sqrt{r}$. Thus the place where p_1 lands on L will give us a coordinate of the desired value. Bingo.

---

[1]Technically, we would want to consider the paper to be the complex plane $\mathbb{C}$ if we were doing a strict algebraic approach. Again, see [Alp00].

Handout 2 (solution and pedagogy)

Simulating this activity in Geogebra or Geometer's Sketchpad should be very straightforward for anyone familiar with the software. If you do not have such software available for your students, I highly recommend exploring Geogebra, since anyone can download it for free at http://www.geogebra.org.

The handout leaves things "blank" in order to make the students think about what they're doing as they do it. You should feel free to just tell students to model the folding activity on geometry software without the handout as a guide if you have the time for students to figure it out for themselves (which will make them more likely to understand it all).

The "fill-in-the-blank" answers are: When we fold a point p to a point p', the crease line we make will be the *perpendicular bisector* of the line segment $\overline{pp'}$.

Then, the steps to finish the construction, after steps (1)–(3) on the handout, are to

(1) draw a line segment between p and p',

(2) draw the midpoint of the segment $\overline{pp'}$,

(3) and then draw a line through this midpoint that is perpendicular to $\overline{pp'}$. (That is, draw the perpendicular bisector of $\overline{pp'}$.)

That perpendicular line is the crease line. With only this line selected, turn on the **Trace** feature (in Geogebra, you do this by CTRL-clicking or right-clicking on the line and selecting "Show Trace") and then move the point p' back and forth along L. Something like the below picture should result.

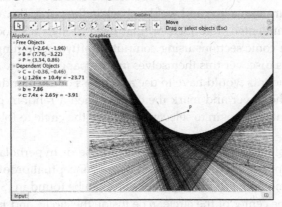

The fun thing about such dynamic geometry software is that as you construct these points and lines, they remain linked. So students should explore moving the point p around, refreshing the trace lines, and seeing how this changes the parabola.

Even better is to let students use the **Locus** command to actually draw the parabola. In Geogebra, make a line through p' that is perpendicular to L. Construct a point q where this line intersects the crease line. Then select the Locus tool (in the "perpendicular line" pop-down menu), click on the point q, and then click on the point p'. This will make Geogebra draw the locus of the point q as p' is moved back and forth along L. And since q is the point of tangency of our line to the parabola, this will result in the parabola being drawn. Then it can be a lot of fun to move the point p around and see how the parabola changes.

The follow-up activity is a *must*. The construction is the same as for the parabola, but start with a circle instead of a line L. The resulting picture will depend entirely on whether or not the students put the point p' inside or outside the circle. Be sure to listen for exclamations of excitement and awe as students develop pictures such as those below.

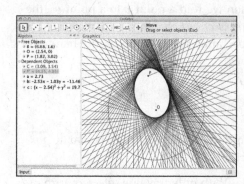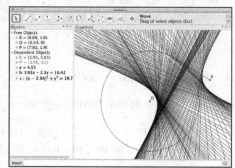

If p' is inside the circle, we get an ellipse with foci p and the center of the circle. If p' is outside the circle we get a hyperbola with the same foci. At a higher concept level, this makes perfect sense—if we transform the center of the circle to infinity, the circle would turn into a line and we'd be back in the parabola case.

Some have commented, like [Sch96], that even though it's cool, quick, and easy to make such conic sections using computer software, nothing compares to letting the students discover this themselves *first* by paper folding. For the ellipse and hyperbola, students would have to use a compass or circular drawing tool to draw a circle on the paper and mark the center. Then a random point p can be chosen, and students can begin to select points p' on the circle to fold to p, unfold, and repeat.

Proving that folding with a circle gives an ellipse or hyperbola is a bit more involved than the parabola case. I'll present here a conceptual proof of the elliptic case. An analytic method and the hyperbola case can be found in [Smi03].

Let O denote the center of the circle, p be inside the circle, and p' be any point on the circle. Then, our crease will be the perpendicular bisector of $\overline{pp'}$, and let x be the intersection point of the crease line and the segment $\overline{Op'}$. Now, recall that an ellipse is determined by two foci and a fixed length l, where the sum of the distances between any point on the ellipse and the two foci is always l.

Claim: The crease line is tangent to the ellipse whose foci are O and p and whose fixed length is the radius of the circle.

Proof: First we show that the crease line contains a point on this ellipse. Since the radius of the circle equals $Ox + xp'$ and $px = xp'$ (by the folding), we know that $Ox + px =$ the radius of the circle, which means that the point x is on the ellipse. Figure (a) below illustrates this.

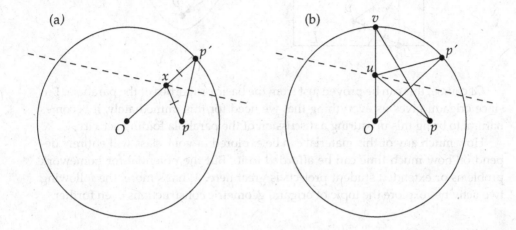

Now we want to show that no other point on the crease line can be on the ellipse, thus proving tangency. Let u be another point on the crease line, and suppose that u *is* on the ellipse. Let Ov be the radius line that contains u. (See figure (b) above.) Since u is on the ellipse, we have that $up = uv$. (Yes, this is clearly not true in the figure, but keep reading.) But since u is on the crease made from folding p' to p, we know that $up = up'$ as well, so $uv = up'$. Since $Ou + up' =$ the radius of the circle, which equals Op', we must have that u lies on the line Op'. This means that $u = x$, and we have that x is the only point on the crease line that is tangent to the ellipse.   □

Final thoughts

You may have noticed that there is a lot of material that can be explored in this activity. Indeed, there's more than what has been touched upon here.

As one last example, in [Smi03] Scott G. Smith mentions how the parabola folding activity gives a nice "proof by origami" that parabolic mirrors have interesting reflective properties. In the following figure, where p is folded onto p' so that x is the point where the crease line is tangent to the parabola, notice that congruent triangles and vertical angles give us that angles α and θ are equal. Thus we can think of α as the angle of incidence and θ as the angle of reflection (or vice versa) of light or sound waves either coming into the parabola and meeting at p or emanating from p and reflecting off the parabola in parallel directions. This is why parabolic surfaces are used for spotlights, stereo speakers, and satellite dishes.

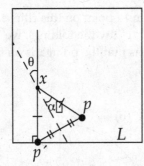 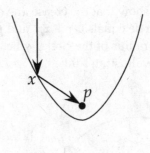

Of course, this can be proven just from the basic properties of the parabola. But since origami gives us everything that we need for this immediately, it becomes natural to bring this up during a discussion of the parabola folding activity.

How much any of this material can be explored in your class will entirely depend on how much time can be afforded to it. But the potential for homework problems or extended student projects is great here. What's more, the following two activities explore the topic of origami geometric constructions even further.

Activity 7
CAN ORIGAMI TRISECT AN ANGLE?

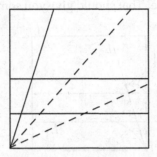

For courses: geometry, abstract algebra

Summary

Students are shown a paper folding routine that seems to trisect any acute angle. Is it for real? A proof or refutation is needed.

Content

The heart of this activity is straightforward geometry. However, the implications from the fact that origami can trisect angles are, for one, that origami is a more powerful construction method than straightedge and compass. This means that the field of origami constructible numbers is larger than the smallest subfield of $\mathbb{C}$ closed under square roots. (See the previous activity for a lead-in to this.)

This activity can be especially captivating in the context of a discussion on the classic Greek problems of trisecting an angle and doubling the cube.

Handout

There is only one handout, which leads students through the angle trisection method and asks them to figure out what it is doing and how to prove it.

Time commitment

The folding part of the activity will take 10–20 minutes of class time, but proving it will take much longer for the students to figure out themselves. Feel free to assign the actual proof for homework.

What's This Doing?

Take a square piece of paper and fold a line from the lower-left corner going up at some angle, θ. Then fold the paper in half from top to bottom and unfold. Then fold the bottom 1/4 crease line. That should give you something like the left figure below.

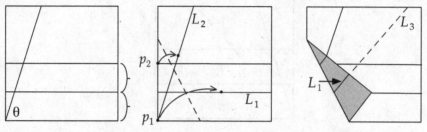

Then do the operation in the middle figure: Make a fold that places point p_1 onto line L_1 **and at the same time** places point p_2 onto line L_2. You will have to curl the paper over, line up the points, and then flatten.

Lastly, with the flap folded, extend the L_1 crease line shown in the right-most figure. Call this crease line L_3.

Question 1: Unfold everything. Prove that we if we extend L_3 then it will hit the lower-left corner, p_1.

Question 2: What is crease line L_3 in relation to the other lines in the paper? Can you prove it, or is this just a coincidence?

SOLUTION AND PEDAGOGY

Handout: Angle trisection

Yes, this routine is showing how one can trisect an acute angle via paper folding. After doing the routine, unfolding everything, and extending L_3 to reach the point p_1, fold the bottom edge of the paper up to L_3 (thus bisecting the angle between L_3 and the bottom side of the square). Then, depending on how accurate the folds were, one can see that angle θ has been trisected.

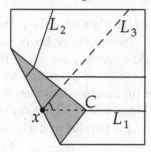 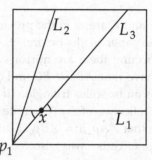

Question 1. The above picture shows why p_1 lies on line L_3. If we let x be the left endpoint of the *segment* of L_3 formed by the folding, we see that the segment p_1x is the same as xC when the paper is folded. The angle between xC and L_3 on the folded paper is thus the same as both angles shown around the point x above on the unfolded paper. Thus the vertical angles around x are equal, and p_1x forms a straight line with L_3.

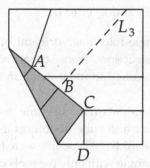 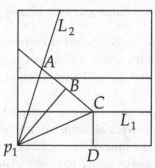

Question 2. There are several ways to prove that this routine is a valid angle trisection. The most simple uses the above figures. We let points A, B, and C be the images of the three points on the left side of the paper after the fold. We also drop a perpendicular from C to meet the bottom edge at D. Then by the definition of these points (as in the above figure, left), we have $AB = BC = CD$. Looking at the unfolded paper, we also have that $p_1B \perp AC$. Thus $\triangle ABp_1$, $\triangle BCp_1$, and $\triangle CDp_1$ are congruent right triangles. Thus they trisect angle θ at p_1.

One can find this angle trisection referenced a number of places on the web, but most of those sites use a different proof that involves the following figure.

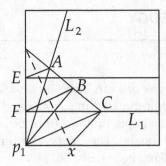

Here A, B, and C are as in the previous proof, and we label the points on the left side of the paper p_1 (as before), F, and E. These two triples of points and the lines connecting them are mirror-symmetric to each other about the crease line. So we have that $AB = BC$ and $p_1B \perp AC$. This is enough to show us that $\triangle p_1AC$ is an isosceles triangle. (If further proof is needed, then notice that $\triangle p_1AC$ is the reflection of $\triangle p_1CE$ under the folding, which is certainly isosceles.) Thus we have that $\angle Ap_1B = \angle Bp_1C$, and since L_1 is parallel to the bottom edge, $\angle FCp_1 = \angle Cp_1x$. Thus θ was trisected.

Pedagogy

This angle trisection method was developed by Hisashi Abe and published in 1980 [Hus80]. There are others, like Jacques Justin's [Bri84]. All have as their fundamental origami "move" the following:

> Given two points p_1 and p_2 and two lines L_1 and L_2, we can make a crease that simultaneously places p_1 onto L_1 and p_2 onto L_2.

This turns out to be the most complicated single-fold basic origami operation possible, and it's what separates origami constructions from straightedge and compass constructions. This operation will be studied in more detail in the next activity.

This activity won't have any impact on the students without some discussion about the controversial history of angle trisections and cube-doublings in mathematics. From the time of the ancient Greeks to the mid-1800s people were trying to develop a means by which to trisect an arbitrary angle with only the tools of an unmarked straightedge and a compass. (Note that people as far back as Archimedes knew that if we used a *marked* straightedge, we could achieve angle trisections. See [Mar98].) Then mathematicians finally proved that angle trisection was impossible with these tools, and in general one can use Galois Theory to prove that an unmarked straightedge and compass cannot solve cubic equations in general.

Now, the mathematical world is full of "false proofs" that straightedge and compass can trisect angles. It is not uncommon for geometry experts to receive letters and emails from amateur mathematicians who claim to have "solved" the problem of angle trisection by the Greek methods. Of course, all such attempts contain some flaw. Often they are very clever and seem to come *very close* to trisecting the angle. But actually doing it perfectly for all angles with straightedge and compass is impossible. Thus, no student of mathematics should accept an origami angle trisection without a rigorous proof that it works!

Without an appreciation of this, these activities will seem rather pointless. But in the context of a geometry, algebra, or history of math class, this can be a real eye-opener as well as serve as a way to solidify exactly what the controversy about angle trisections was for all those years. That is, seeing an easy way how one could trisect angles with origami helps one understand why it couldn't be done with other tools. Such an understanding would be achieved more readily using the next activity on cubic equations in addition to this one.

Since there is more than one way to prove the trisection, you should let students play with trying to prove it on there own for a good amount of time before giving any hints. It often doesn't occur to students that they should draw the image of the left side of the paper (line AC in the figures), so this can be a gentle suggestion that doesn't give everything away. In fact, any proof will be very difficult to develop without an understanding of how when we fold paper, part of the paper is *reflected* about the crease line, and thus lengths and angles are preserved under this transformation. That "creases are reflections" is a fundamental ingredient of these proofs, and discussion and/or demonstration of this beforehand may be very useful. (In fact, this activity can be a great way to reinforce concepts of reflection transformations.)

Follow-up

If your students take to this activity, you might want to let them see how a similar "two points to two lines" fold can solve another classic Greek problem: doubling the cube. This problem asks one to construct a cube that is twice the volume of a given cube, and this is equivalent to constructing $\sqrt[3]{2}$. Again, straightedge and compass cannot perform this task, but origami can.

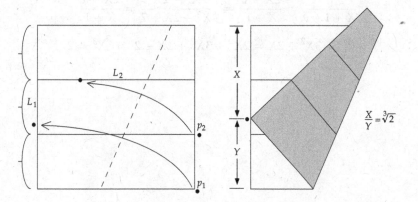

The following method was developed by Peter Messer [Mes86]. First a square piece of paper needs to be folded into thirds; see the Dividing a Length into *n*ths

Exactly activity for instructions on how to do this. Then, using the points and lines as labeled in the previous figure, do the origami "move" of folding p_1 onto L_1 and p_2 onto L_2 simultaneously. The image of p_1 under this fold will divide the left side of the paper into two lengths, the ratio of which is $\sqrt[3]{2}$.

Proving that this works is a very challenging Euclidean geometry exercise. None of the steps are particularly hard, but the elements of this problem have a tendency to get out of control, generating overly complicated equations unless done in the proper sequence. A helpful trick is to let $Y = 1$, so that the side of the square is $X + 1$. Then all we need to do is prove that $X = \sqrt[3]{2}$.

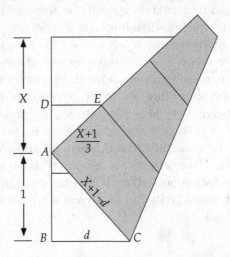

Label things as in the above figure. We can use the Pythagorean Theorem on $\triangle ABC$ to get that $d = (x^2 + 2x)/(2x + 2)$. Also, the length of AD is $X - (X + 1)/3 = (2X - 1)/3$. Now, $\triangle ABC$ and $\triangle ADE$ are similar (see the Haga's "Origamics" activity for details, as this is just Haga's Theorem), so we have

$$\frac{d}{X + 1 - d} = \frac{2X - 1}{X + 1} \Rightarrow \frac{X^2 + 2X}{X^2 + 2X + 2} = \frac{2X - 1}{X + 1}$$

$$\Rightarrow X^3 + 3X^2 + 2X = 2X^3 + 3X^2 + 2X - 2 \Rightarrow X^3 = 2.$$

Bingo!

Activity 8
SOLVING CUBIC EQUATIONS

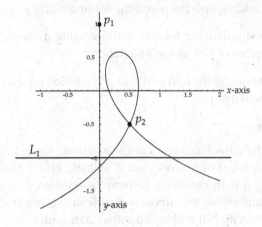

For courses: geometry, abstract algebra

Summary

Students are led through a paper folding activity that explores the type of equations that are generated by a more powerful origami operation. The activity involves folding the paper and drawing points that, when connected, form strange-looking curves. Students model this process and discover that the curve generated is actually made by a cubic equation. These curves can also be generated using dynamic geometry software.

Content

This is a more advanced foray into geometric constructions via origami. As such, it wouldn't be feasible to do this activity without first doing the previous two (on folding a parabola and trisecting an angle) with the class. In fact, the origami operation that this activity explores is exactly the key step in the angle trisection construction. Without that as motivation it would be difficult to get students to understand the significance of the "two points to two lines" fold.

This folding operation has both geometric and algebraic interpretations. Geometrically, it's equivalent to finding a common tangent line to two parabolas drawn in the plane. Algebraically, it's equivalent to solving a general cubic equation. Either of these can be explored in depth, depending on the focus of one's course.

69

Handouts

Because of the need to motivate this folding operation, the handouts assume that the students have done the angle trisection and parabola exercises.

(1) Introduces the folding operation and the folding activity.

(2) Helps students simulate the folding activity using dynamic geometry software, like Geogebra or Geometer's Sketchpad.

(3) Asks students to model the fold and find an equation for the curve generated in the folding activity.

Time commitment

Students who have done the Folding a Parabola activity should have no problem with the folding component of this one, but it will still take a good 20 minutes of class time. Modeling it with geometry software should only require 10–15 minutes. Deriving the equation for the curves is doable in class by students who completed the parabola activity, but will take another 20 minutes.

A More Complicated Fold

The origami angle trisection method is able to do what it does by using a rather complex origami move:

> Given two points p_1 and p_2 and two lines L_1 and L_2, we can make a crease that simultaneously places p_1 onto L_1 and p_2 onto L_2.

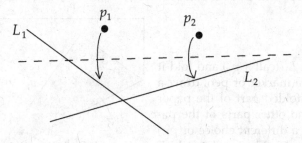

Question 1: Will this operation always be possible to do, no matter what the choice of the points and lines are?

Question 2: Remember that when we fold a point p to a line L over and over again, we can interpret the creases as being tangent to a parabola with focus p and directrix L. What does this tell us about this more complex folding operation? How can we interpret it geometrically? Draw a picture of this.

Activity: Let's explore what this operation is doing in a different way. Take a sheet of paper and mark a point p_1 (somewhere near the center is usually best) and let the bottom edge be the line L_1.

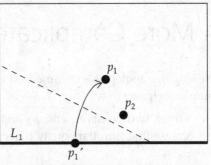

Pick a second point p_2 to be anywhere else on the paper. Our objective is to see where p_2 goes as we fold p_1 onto L_1 over and over again.

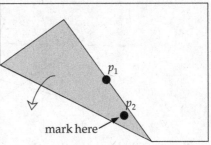

So pick a spot on L_1 (call it p_1') and fold it up to p_1. Using a marker or pen, draw a point where the folded part of the paper touches p_2. (If no other parts of the paper touch p_2, try a different choice of p_1'.) Then unfold. You should see a dot (which we could call p_2') that represents where p_2 went as we make the fold.

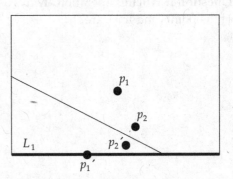

Now choose a different p_1' and do this over and over again. Make enough p_2' points so that you can connect the dots and see what kind of curve you get.

Question 3: What does this curve look like? Look at other people's work in the class. Do their curves look like yours? Do you know what kind of equation would generate such a curve?

Simulating This Curve with Software

We're still considering this unusual origami maneuver:

> Given two points p_1 and p_2 and two lines L_1 and L_2, we can make a crease that simultaneously places p_1 onto L_1 and p_2 onto L_2.

So that you don't have to keep folding paper over and over again, let's model our folding activity using geometry software, like Geogebra. This will allow us to look at many examples of the curve this operation generates and do so very quickly.

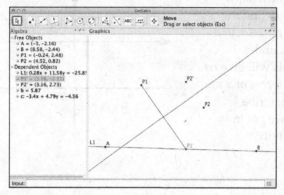

Here's how to set it up:

(1) Make the line L_1 and the point p_1.

(2) Make a point p_1' on L_1 and construct a line segment from p_1 to p_1'.

(3) Construct the perpendicular bisector of $\overline{p_1 p_1'}$. This makes the crease line.

(4) Now make a new point, p_2.

(5) Reflect the point p_2 about the crease line made in step (3). In Geogebra, this is done using the **Reflect Object about Line** tool. The new point should be labeled p_2'.

Then when you move p_1' back and forth along L_1, the software will trace out how p_2' changes. You can either draw this curve by turning on the **Trace** of p_2' (CTRL-click or right-click on p_2' to turn this on in Geogebra) or use a **Locus** tool to plot the locus of p_2' as p_1 changes.

Activity: Move p_2 to different places on the screen and see how the curve changes. How many different basic shapes can this curve take on? Describe them in words.

What Kind of Curve Is It?

To see what type of curve this operation is giving us, make a model of the fold.

Let $p_1 = (0, 1)$.
Let L_1 be the line $y = -1$.
We'll fold p_1 to $p_1' = (t, -1)$ on L_1.
Let $p_2 = (a, b)$ be fixed.
Then, we want to find the coordinates of $p_2' = (x, y)$, the image of p_2 under the folding. This will give us an equation in terms of x and y that should describe the curve that we got in our folding activity.

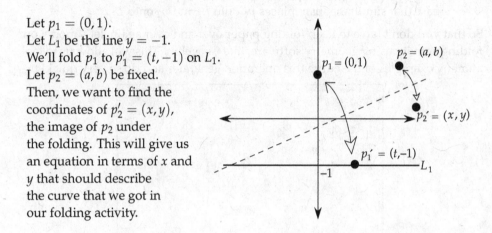

Instructions: Find the equation of the crease line that we get when folding p_1 onto p_1'. Use this and the geometry of the fold to get equations involving x and y. Combine these to get a single equation in terms of x and y (with the constants a and b in it as well, but no t variables). What kind of equation is this?

SOLUTION AND PEDAGOGY

Handout 1: A more complicated fold

This may be the first time students have seen this folding operation stated explicitly. It would be useful to compare this statement to the angle trisection instructions, if only to convince students that this fold can be done.

Question 1. No, this fold is not always possible to do. If one imagines the lines L_1 and L_2 to be parallel and far apart and p_1 and p_2 close together in between the two lines, one can see that putting p_1 on L_1 and p_2 on L_2 would be impossible. (After all, every fold is an isometry, so the distance between p_1 and p_2 has to be preserved.)

Question 2. Since we're folding p_1 onto L_1, the crease line that we make will be tangent to the parabola with focus p_1 and directrix L_1. Similarly, the crease line will also be tangent to the parabola with focus p_2 and directrix L_2.

   Therefore, this folding operation is equivalent to finding a common tangent line to two different parabolas.

Folding activity. Like the Folding a Parabola activity, this one requires many folds and many plotted points p_2' to generate a reasonably good curve. As more folds are made, there often comes a time when the fold actually *moves* the point p_2 instead of just bringing a layer of paper on top of it. In these cases a mark must still be made in the paper where p_2 goes.

   The below figure shows one possible example, with x- and y-axes shown for reference.

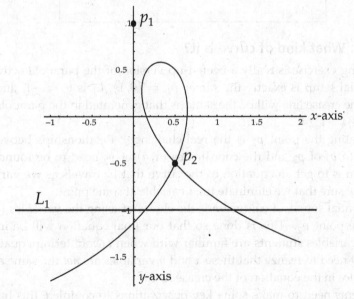

Question 3. The curve should look like a cubic equation; however, it's very likely that students may never have seen the graph of a genuine cubic equation before, so they may not be able to conjecture what it is.

Handout 2: Simulating this curve with software

After the paper folding activity would be a good time to have the students explore more such curves on Geogebra, if the computer resources needed for this are at hand. Again, nothing compares to the students plotting these curves themselves by actually folding paper, but Geogebra will allow each student to experience the variety of shapes that cubic curves can create.

If the students are experienced with software like Geogebra, you can have them develop this simulation without the handout. But using the **Reflect**, **Trace**, and **Locus** features might not be familiar to some students, so detailed instructions on how to set this up are included on the handout. Below is a sample screen shot of what students might see.

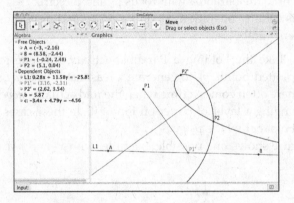

Handout 3: What kind of curve is it?

This modeling exercise is really a beefed-up version of the parabola activity. In fact, the initial setup is exactly the same: $p_1 = (0, 1)$, L_1 is $y = -1$, and $p_1' = (t, -1)$. So the crease line will be the same as that generated in the parabola activity: $y = (t/2)x - t^2/4$.

Incorporating the point p_2 is the real challenge. Relationships between the coordinates (a, b) of p_2 and the coordinates (x, y) of p_2' need to be found. And since our aim is to get an equation of the curve that p_2' travels as we vary t, we want to make sure that we eliminate the t variable at some point.

One potential source of confusion is the choice of using the variables (x, y) to represent the point p_2'. This is done so that our final equation will be in terms of x and y, variables students are familiar with when encountering equations of curves. They need to realize that these x and y variables are *not* the same as the x and y variables in the equation of the crease line.

So, students need to make some key observations to complete this handout. First, the slope of the line segment $\overline{p_1 p_1'}$ should be the same as that of $\overline{p_2 p_2'}$. This means that

$$-\frac{2}{t} = \frac{y-b}{x-a}. \tag{1}$$

Many students will then be tempted to plug this into the equation of the crease line to obtain a single equation with only x and y as variables. (Remember, a and b are constant.) But this is flawed because the x, y variables in the crease line are not the same as those of p_2'. In fact, some students may choose to label the p_2' coordinates as (x', y') just to distinguish them.

However, there is a point that we know is on the crease line—the midpoint of $\overline{p_2 p_2'}$, which is $((x+a)/2, (y+b)/2)$. If we plug this point into the crease line equation, and then plug (1) in for the $t/2$ variables, we obtain a valid equation in terms of the $p_2' = (x, y)$ coordinates:

$$\frac{y+b}{2} = -\left(\frac{x-a}{y-b}\right)\frac{x+a}{2} - \frac{(x-a)^2}{(y-b)^2}$$

$$\Rightarrow (y+b)(y-b)^2 = -(x^2 - a^2)(y-b) - 2(x-a)^2.$$

Notice that this is a cubic curve! (We have a y^3 term on the left-hand side and an $x^2 y$ term on the right.)

Unfortunately, seeing this equation might not be as fundamentally thrilling to a typical undergraduate math major as it would be to faculty. But plotting this equation for specific values of (a, b) can be very illuminating, as it generates the same curves that the students were creating with the folding activity. (For example, the plot in the folding activity section was made using $(a, b) = (.5, -.5)$.) Plotting such an equation requires either Maple, Mathematica, or an expensive enough graphing calculator, but making students do this is *very* worthwhile.

Pedagogy and follow-up

As stated previously, this exercise is an advanced paper folding activity, and it should only be investigated after the previous two activities on folding a parabola and trisecting an angle. In this way, the parabola activity combines paper folding with a subject—parabolas, conic sections, and their equations—with which students are already familiar. Then students will likely have heard of angle trisections before, so that activity will also be combining the familiar with the novel. This is then carried into the current activity, where everything is likely to be completely new to the students. The effect, from the students' perspective, can be one of being brought into much more advanced mathematics with a much "deeper" feel. Indeed, the fact that cubic equations cannot be so easily classified and are unfamiliar to students can create such a feeling.

Thus, in the folding exercise in this activity, students are wading into unfamiliar waters. They'll need to understand the "rules" of the activity very clearly before embarking. Also, it helps to make sure that students in the class sample a wide variety of choices for the point p_2. Some will produce "loops" like the sample plot previously given. Others will seem to have a "cusp" point or be an ordinary-looking curve with an odd bump in it. In fact, instructors might want to have all students start with the same choice for p_1 and then make sure a good enough sampling of p_2 points are chosen so that a wide enough array of cubic curves will be seen.

Using the model in the third handout to create the equation for the cubic curve can be very difficult for students. This is a great example of something that seems very easy once one sees how to do it but beforehand seems incredibly difficult. It's also hard for students to see the approach of this model, which is fundamentally different from our approach to the parabola problem. For the parabola we wanted to find the curve tangent to all the crease lines. Now we're trying to study the *behavior* of p_2' as we fold p_1 to L_1 over and over again. Thus we want some equation involving only the variables x and y (and not t, although a and b are okay because they're constants) that reflects what p_2' is doing as we fold. And *any* such equation that we generate from the model should do the trick (provided it isn't overly complicated), since it would give us a constraint on the possible coordinates of p_2'.

Note that in the folding activity we ignore completely the role that line L_2 plays in this origami move. The justification for this is because if we folded p_2 to a line L_2, then the fold would be determined by folding p_2 to a spot where L_2 intersects our cubic curve. Locating such a spot would be "solving" our equation at a specific point.

As stated above, it's incredibly useful for the students to be able to plot these cubic equations on a computer or graphing calculator to see directly that they look like the curves generated by the folding activity. Being able to connect the mathematical model to the physical activity can be a great moment of clarity for students.

Pesky question. As in the parabola activity, this one gives evidence that origami can solve certain equations, in this case cubic equations, but proving this would require giving an argument that an arbitrary cubic equation can always be solved via origami. So how do we do this?

There are several different methods one could use to employ origami to solve an arbitrary cubic equation. In what follows I will describe a method due to Alperin, found in [Alp00]. However, while this method is technically correct, it is not nearly what most students could conceive of creating themselves. For a different, more explicitly hands-on cubic-solving method, see the next activity on Lill's Method.

For Alperin's strategy, one would like to start with an arbitrary cubic of the form $x^3 + ax^2 + bx + c = 0$, but we can actually get rid of the x^2 term by substituting $z = x - (1/3)a$. This gives us

$$z^3 + \frac{3b - a^2}{3} z - \frac{9ab - 27c - 2a^3}{27} = 0.$$

Thus we can actually assume that our general cubic is of the form $x^3 + ax + b = 0$, where a and b are rational. Consider the quadratic equations

$$\left(y - \frac{1}{2}a\right)^2 = 2bx \quad \text{and} \quad y = \frac{1}{2}x^2.$$

Since a and b are constructible via origami, the coefficients of these two equations can be constructed, and thus so can the foci and directrices of these two parabolas. The first parabola will have focus $(b/2, a/2)$ and directrix $x = -b/2$, and the second will have focus $(0, 1/2)$ and directrix $y = -1/2$. (These can be computed using standard precalculus formulas, which of course most people will have forgotten but can look up in any precalculus text.)

So, fold $(b/2, a/2)$ onto $x = -b/2$ and $(0, 1/2)$ onto $y = -1/2$ simultaneously. This will produce a crease that is tangent to both of these two parabolas, and while this folding move can sometimes have more than one possible fold, in this case our crease line is unique. Let m be the slope of this crease line.

Claim: m is a root of $x^3 + ax + b = 0$.

Proof: Let (x_0, y_0) be the point of tangency of the crease line with the first parabola and (x_1, y_1) be the tangent point with the second parabola. We can take derivatives of the two parabolas and plug in these points and m to yield some equations. The first parabola gives

$$2\left(y - \frac{a}{2}\right)\frac{dy}{dx} = 2b \Rightarrow m = \frac{b}{y_0 - a/2}.$$

The second yields $m = x_1$, and so $y_1 = (1/2)m^2$. Also, plugging (x_0, y_0) into the first parabola equation gives $x_0 = (y_0 - a/2)^2/(2b) = b/(2m^2)$. However, m can also be computed in the traditional way:

$$m = \frac{y_1 - y_0}{x_1 - x_0} = \frac{\frac{m^2}{2} - \frac{a}{2} - \frac{b}{m}}{m - \frac{b}{2m^2}}.$$

Simplifying this, amazingly enough, gives $m^3 + am + b = 0$. Wow. $\qquad \square$

Since we've created a crease line with slope equal to a real root of our arbitrary cubic, we can easily construct a coordinate with m in it. For example, if we let $(w, 0)$ be the point where the crease line crosses the x-axis (which is thus a constructible point), then we fold the line $x = w + 1$. This vertical fold will intersect the crease line at the point $(w + 1, m)$. Then we can officially say that we've constructed a root of our cubic via origami.

Other questions. There are several follow-up questions that can be asked, which may make good homework questions and such.

One is, "Could the origami move under consideration have more than one possibility?" The answer is yes, sometimes. Algebraically this makes sense because we now see that it is equivalent to solving a cubic, which may have as many as three real solutions. But it can also, and perhaps more readily, be seen graphically. Since the fold can be determined by folding p_2 to a point where L_2 crosses our cubic equation curve, we really need to determine the number of intersection points that can be possible. Familiarizing yourself with the shape of cubic curves will convince you that at most three such intersection points can occur between a cubic curve and a straight line, as can two or one or none.

Another question would be to ask if the "cubic" origami move is really different from the move encountered in the Folding a Parabola activity. Actually, the latter move is a special case of the former. If the point p_2 is already on the line L_2, then "folding p_2 to L_2" is really just folding p_2 to itself, which is tantamount to making sure the crease goes through p_2. It can be interesting to see how other, more simple origami operations can be viewed as special cases of the cubic move.

Abstract algebra approach. One can approach all this from an algebraic perspective, which is the more rigorous way to do it, after all. The idea is to analyze origami constructions similarly to how one analyzes straightedge and compass (SE&C) constructions.

When modeling SE&C constructions algebraically, we typically consider the paper that we're drawing on to be the complex plane $\mathbb{C}$ and start off with some given starting points, like the origin, the point 1, and the point i, say. Then we ask, "What subfield of $\mathbb{C}$ can we construct using only the tools of SE&C?" We call this the *field of SE&C constructible numbers*, and it is the smallest subfield of $\mathbb{C}$ that is closed under square roots. In other words, $\alpha \in \mathbb{C}$ is SE&C constructible if and only if α is algebraic over $\mathbb{Q}$ and the degree of its minimal polynomial over $\mathbb{Q}$ is a power of 2, i.e., $[\mathbb{Q}(\alpha) : \mathbb{Q}] = 2^n$ for some integer $n \geq 0$. That is, SE&C can solve quadratic equations.

We can ask the same questions about origami. We think of our sheet of paper as $\mathbb{C}$ and assume that we're given some points to start off with, like the origin, 1, i, and maybe $1 + i$ (to simulate the four corners of a square). We then want to find the subfield $\mathcal{O} \subset \mathbb{C}$ that origami moves can generate from these points. We call this the field of *origami numbers*. There are lots of basic origami moves that we can assume, like given two points, we can make a crease line connecting them, or folding one point to another, or folding a line onto another line. That can get us started to generate the rationals, say.

In the Folding a Parabola activity we saw how the move of folding a point to a line can solve general quadratic equations, and this means $\mathcal{O}$ contains the set of SE&C constructible numbers.

The fold-two-points-to-two-lines origami move studied in this activity, however, shows that $\mathcal{O}$ is bigger than the SE&C subfield. In fact, we proved that $\mathcal{O}$ contains all solutions to cubic equations over the rationals. Stating this more formally takes more work, or rather *proving* it takes more work. One way to restate it is the following.

Theorem: *Let $\alpha \in \mathbb{C}$ be algebraic over $\mathbb{Q}$ and let $L \supset \mathbb{Q}$ be the splitting field of the minimal polynomial of α over $\mathbb{Q}$. Then α is an origami number if and only if $[L : \mathbb{Q}] = 2^a 3^b$ for some integers $a, b \geq 0$.*

The proof is done basically by formalizing, using field extensions, what we do when we use either a point-to-line fold (requiring a quadratic) or a two-point-to-two-lines fold (requiring a cubic) to generate more and more points. See [Cox04] for an excellent description of how to do all this. Also see [Mar98] and [Alp00].

Note, however, that the theorem above assumes that the fold-two-points-to-two-lines origami move is the most complicated move that we can make. After seeing this move, students might very well wonder whether or not there are any other, more complicated origami moves that are possible. This turns out to be a very complicated question. If we assume that all our creases are straight lines and that we are allowed to make only one fold at a time, then it can be proven that the fold-two-points-to-two-lines move is the most complex move possible (in terms of the highest degree equations that it is capable of solving in general). This was proven by Robert Lang using vector analysis in 2003 [Lang03], and a shorter, geometric version of this proof can be found in [Hull05-1].

If we deviate from these restrictions, however, more is possible. Robert Lang discovered an ingenious way to *quintisect* an arbitrary angle by incorporating a very complicated maneuver that requires making two creases simultaneously [Lang04-2]. Angle quintisections require the solving of fifth-degree equations.

If you show Lang's angle quintisection method to your students, you should let them debate whether or not such origami moves should be allowable. How is Lang's two-simultaneous-creases move different from folding a sheet of paper into perfect thirds lengthwise by making the two creases simultaneously? When do such "simultaneous creases" origami moves become too complicated for humans to handle? These are all questions that are still being debated in the origami mathematics community and can make for lively classroom discussions as well.

Activity 9
LILL'S METHOD

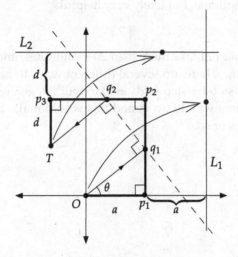

For courses: geometry, abstract algebra

Summary

Building on the Solving Cubic Equations activity, students are introduced to, and asked to prove, the cubic equation case of Lill's Method, a geometric construction technique for solving polynomial equations. This construction method is possible to do via origami, and students are asked to test it out to find the roots, using only paper folding, of a given cubic polynomial.

Content

Despite the inherent difficulty of solving cubic equations, Lill's Method is fairly simple. In fact, proving that it works only requires basic trigonometry and algebra. The folding needed to do this is precise but not difficult.

The deeper concepts underlying this activity would be suitable for any geometry class. That is, it gives an understandable, constructive proof that origami can solve any cubic equation. The algebra-geometry connection demonstrated here would be suitable for an abstract algebra class covering constructible numbers.

Handouts

There are three handouts for this activity:

(1) Describes Lill's Method for cubics and asks students to prove that it works.

83

(2) Describes how to do Lill's Method for cubics via origami and asks students to try it for a given (nice) cubic polynomial.

(3) Gives step-by-step folds for setting up the activity in the second handout. This handout is optional, but likely very helpful.

Time commitment

The first handout should not take more than 20–30 minutes, and it could even be assigned for homework. There are several different ways to handle Handouts 2 and 3. Going over the step-by-step folds in Handout 3 together as a class would be the most time-efficient way to do it, and this would make Handouts 2 and 3 take another 20 minutes or so.

Lill's Method for Solving Cubics

In this activity you'll learn Lill's Method for using geometry to solve cubic equations. Lill's Method is cool because we can do it via origami!

Imagine we want to solve (find a real root) of the following cubic:

$$ax^3 + bx^2 + cx + d = 0.$$

Setup: Start at the origin, point O, and draw a line segment of length a along the positive x-axis. Then rotate $90°$ counterclockwise and go up a length of b. Repeat: Turn $90°$ counterclockwise and go a length of c, then turn once more and travel a distance d, ending at a point T.

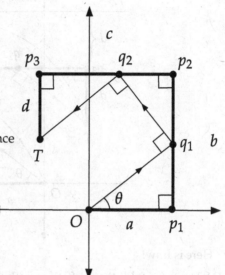

Note: If any of the coefficients are negative, then go backwards. If any are zero, then rotate but do not travel.

Then imagine that we stand at the point O and try to "shoot" T with a bullet that bounces off the coefficient path at right angles, as shown.

Lill's Method states that if we can successfully hit the point T with such a bullet path, and θ is the angle the path makes at O, then $x = -\tan\theta$ will be a root of our cubic equation!

Your task: Prove that Lill's Method for solving a cubic equation is correct. (Hint: What do you notice about the triangles in the figure? And what is $\tan\theta$ equal to?)

Lill's Method via Origami

We can use origami to solve any cubic equation by using Lill's Method. This idea was discovered by the Italian mathematician Margherita Beloch in 1933.

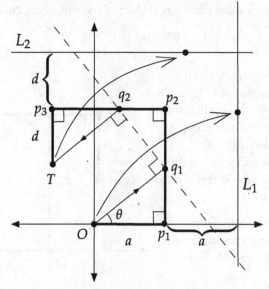

Here is how:

(1) Draw (or fold) the coefficient path from Lill's Method on your paper.

(2) Then fold a line perpendicular to the x-axis at a distance of a from $\overline{p_1 p_2}$ (on the side opposite of O). Call this line L_1.

(3) Then fold a line perpendicular to the y-axis at a distance d from $\overline{p_2 p_3}$ (on the side opposite of T). Call this line L_2.

(4) Then fold O onto L_1 while at the same time folding T onto L_2. This crease line will form one side of the bullet path needed for Lill's Method (and contain the angle θ that we need).

Activity: Study the above instructions, and then try it yourself to find the roots of the polynomial $x^3 - 7x - 6$ only using paper folding.

Doing this will require you to think of the paper as the xy-plane, decide where the origin should be, and fold the coefficient path. Try setting this up as shown to the right, and then fold the point O onto the line L_1 while also placing T onto L_2.

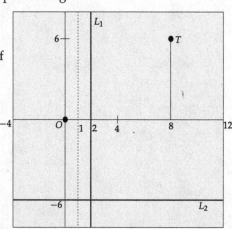

Lill's Method Example: Step-by-Step Folds

This handout provides step-by-step folds for making the crease lines needed to solve $x^3 - 7x - 6 = 0$ using Lill's Method. Begin with a large square piece of paper, and imagine that it goes from $-4 \le x \le 12$ and $-8 \le y \le 8$ in the xy-plane.

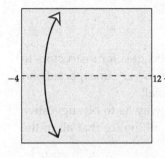

(1) Crease in half to make the x-axis.

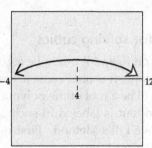

(2) Pinch in half at $x = 4$.

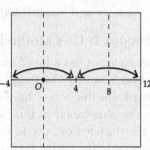

(3) Fold the y-axis and the $x = 8$ line.

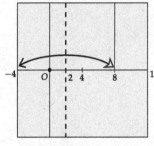

(4) Fold the $x = 2$ line.

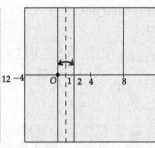

(5) Fold the $x = 1$ line.

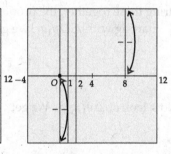

(6) Make these pinches.

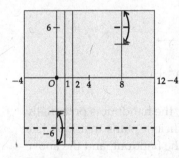

(4) Fold the $y = -6$ line and a pinch at $y = 6$.

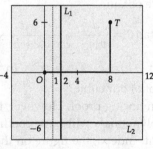

(5) Ready for Lill's Method!

Draw points $O = (0,0)$ and $T = (8,6)$ and lines $L_1 : y = -6$ and $L_2 : x = 2$. The coefficient path is drawn from O to T in bold. Do you see why this is the path?

SOLUTION AND PEDAGOGY

Lill's Method is actually a general, geometric construction that allows one to find real roots of any single-variable polynomial equation. It was discovered by an Austrian engineer named Eduard Lill in the 1800s [Lill1867]. The fact that it can be applied in origami to solve cubic equations was discovered by the Italian mathematician Margherita Beloch in the 1930s [Belo36]. See [Hull11] for more details of this story and this method.

Handout 1: Lill's Method for solving cubics

This handout has a lot of text on it, and it might be a good idea for instructors to go over how Lill's Method is done. It should come across as very unexpected and remarkable that $x = -\tan\theta$ will be a root of the polynomial!

The illustration in this handout is labeled in such a way as to be suggestive to the students of how to prove Lill's Method. First of all, notice that all of the triangles in the figure are similar, and thus

$$\theta = \angle q_1 O p_1 = \angle q_2 q_1 p_2 = \angle T q_2 p_3.$$

Each one of these triangles can, therefore, be used to calculate $\tan\theta$, and if we string these together, the proof falls right out.

Starting with $\triangle O p_1 q_1$, we get

$$-x = \tan\theta = \frac{p_1 q_1}{a} = \frac{b - q_1 p_2}{a}$$

$$\Rightarrow q_1 p_2 = ax + b.$$

Now look at $\triangle q_1 p_2 q_2$. We get

$$-x = \tan\theta = \frac{p_2 q_2}{q_1 p_2} = \frac{c - p_3 q_2}{ax + b}$$

$$\Rightarrow p_3 q_2 = x(ax + b) + c.$$

Finally, looking at $\triangle q_2 p_3 T$, we obtain

$$-x = \tan\theta = \frac{d}{p_3 q_2} = \frac{d}{x(ax + b) + c}$$

$$\Rightarrow 0 = x(x(ax + b) + c) + d = ax^3 + bx^2 + cx + d.$$

Therefore, $x = -\tan\theta$ is a root of our cubic.

This suffices as a fairly convincing proof. However, the handout is potentially misleading. Not all cubic polynomials will have coefficient paths that look like the one shown on the handout. Indeed, the figure on the handout, and the above proof that is based on it, only hold if all the coefficients of the cubic are positive. If any of them are negative or zero, then the coefficient path will look different and affect the algebra in the proof.

As another example, suppose that the coefficient of the x term is negative and the rest of the coefficients are positive. In other words, our cubic is of the form

$$ax^3 + bx^2 - cx + d = 0.$$

Our coefficient and bullet paths will look something like the below figure in this case.

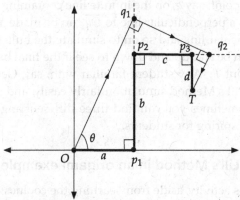

Notice how we need to bounce the bullet through the lines containing segments $\overline{p_1p_2}$ and $\overline{p_2p_3}$ in order to get Lill's Method to work. The algebra in this case will be as follows:

$$\text{In } \triangle Op_1q_1, \ -x = \tan\theta = \frac{p_1q_1}{a} = \frac{b + q_1p_2}{a}$$

$$\Rightarrow \ q_1p_2 = -ax - b.$$

$$\text{In } \triangle q_1p_2q_2, \ -x = \tan\theta = \frac{p_2q_2}{q_1p_2} = \frac{c + p_3q_2}{-ax - b}$$

$$\Rightarrow \ p_3q_2 = -x(-ax - b) - c.$$

$$\text{In } \triangle q_2p_3T, \ -x = \tan\theta = \frac{d}{p_3q_2} = \frac{d}{-x(-ax - b) - c}$$

$$\Rightarrow \ -ax^3 - bx^2 + cx = d \ \Rightarrow \ ax^3 + bx^2 - cx + d = 0.$$

Other cases are handled similarly, including those where one or more of the coefficients are 0. It is certainly too much to expect students to prove *all* of the possible cases, but they should be made aware that the coefficient path pictures can look different and that this changes the proof slightly. (Assigning one of the other cases to prove might make a good homework exercise.)

Lill's Method is nothing short of ingenious. It is surprising that while it seems to have been well-known during the late 1800s and early 1900s, it seems mostly forgotten now. This is beginning to change; recent research on origami geometric constructions has been making use of Lill's Method (see [AlpLan09]).

It is worth noting two further things about Lill's Method. First, Lill's Method works for arbitrary-degree polynomials and thus can find real roots of quintics or any polynomial that you like. The proof is exactly the same. Second, it can be incredibly fun and concept-strengthening to simulate Lill's Method using dynamic geometry software, like Geogebra or Geometer's Sketchpad. To do this, one needs to construct a coefficient path, making sure to have each segment $\overline{p_1p_2}$, $\overline{p_2p_3}$ and so on be contained in an infinite line, since you don't know if the bullet path will need to make use of the extended segments. Then create a potential bullet path by making a movable point, say q, on the infinite line containing $\overline{p_1p_2}$ and making the segment $\overline{Oq}$. Then a perpendicular line to $\overline{Oq}$ can be made at the point q, and then another perpendicular line and so on to simulate the bullet path. When this is done, the point q can be slid up and down to see if the final bullet path line can be made to hit the point T. Any student familiar with, say, Geogebra should be able to create such a Lill's Method simulation fairly easily, and seeing it work (as well as seeing how sometimes you can find three different angles that shoot the point T) can be very inspiring for students.

Handouts 2 and 3: Lill's Method in an origami example

The whole point of this activity, aside from learning the coolness of Lill's Method, is to see in a very hands-on manner how origami can solve arbitrary cubic polynomials. Thus the "Lill's Method via Origami" handout has two purposes: to let students see how paper folding's "fold two points to two lines" move can give us bullet path solutions to Lill's Method when applied to cubics, and then to have students practice it themselves with an example.

Time needs to be spent in class for students to understand what the top half of the second handout is illustrating. The lines L_1 and L_2 are carefully placed so that when we fold O onto L_1 and T onto L_2, the crease line made will contain the second side of the bullet path that we need. (The other two sides of the bullet path can then be folded easily, if desired, by making creases perpendicular to the newly-made crease through q_1 and q_2.) All it takes to convince yourself that this is the case is remembering that if we fold one point to another point, say O to a point O' on L_1, then the crease made will be the perpendicular bisector of $\overline{OO'}$. That gives us the right angles we need for the bullet path.

Folding the example

With luck, students (and faculty) will resist the urge to try to actually factor $x^3 - 7x - 6$, let alone solve it on a computer or calculator. It is a carefully-chosen cubic polynomial that has three integer roots, all of which can be found using Beloch's paper-folding method of solving cubics.

Doing this in practice requires that we set up xy-plane coordinates on our square piece of paper. Any such coordinate range that is big enough will do, but it is better to pick one that works nicely. The coordinate system presented here was developed by Cary Malkiewich while teaching this material with the author during the summer of 2006.

Malkiewich's setup has the square having side length 16 with the x-axis along the horizontal middle going from -4 to 12. This puts the y-axis along the vertical 1/4 line of the paper to the left of the center.

In order to prepare the paper for this coordinate setup, the x- and y-axes need to be folded (in order to locate the origin O), along with the coefficient path (in order to locate the point T) and the lines L_1 and L_2 that are needed for Beloch's strategy for solving Lill's Method. Working out how to make all these folds by yourself can be fun and not too difficult, but people (i.e., faculty and students) without much folding experience might not know where to begin.

Therefore, the third handout is provided to give step-by-step instructions for setting up Malkiewich's coordinate system. Instructors should follow these instructions themselves and then decide if they want to have their own students follow them as well, or if it would be easier for the instructor to simply lead the class through each step.

As the coordinate system is folded into the paper, be sure to draw along the relevant crease lines so as to make it easier to see. This is important for being able to actually do the fold that finds the bullet path. Marking the points at O and the numbers 1, 2, 4, and 8 on the x-axis helps as we do the folding. More importantly, draw along the lines L_1 and L_2 with a thick pen and make large dots at the points O and T.

With the coordinate system in place and the important lines drawn, students will be ready to tackle the main activity of the second handout: trying to find the roots of $x^3 - 7x - 6$. This requires holding up one's square piece of paper and bending it so that O lands on L_1 and T lands on L_2. Holding the paper up to a light so as to see the large dots at O and T through the paper and positioning them on their respective lines can help. But since this polynomial has three real roots, there are three different ways to actually do this, and in a class full of students you will likely see examples of all three ways. One of them is shown below.

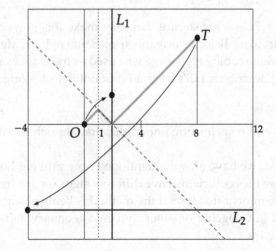

A careful reading of this case shows that the point T gets folded to the point $(-4, -6)$ on L_2 and O gets folded to the point $(2, 2)$ on L_1. This means that the similar triangles made by the bullet path and the coordinate path are all $45°$ right triangles! In other words, the angle θ at O is $45°$, and so $\tan\theta = 1$. Thus $x = -1$ should be a root of $x^3 - 7x - 6$. Checking this, one sees that it is true ($-1 + 7 - 6 = 0$).

The two other possibilities for where O and T can go are shown below.

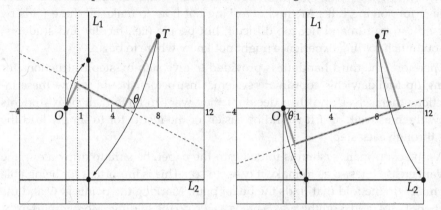

On the left, we see that T folds to the intersection point of L_1 and L_2, which is $(2, -6)$. This means that the crease line intersects the x-axis at $x = 5$ (because the crease is the perpendicular bisector of the segment between $T = (8, 6)$ and $(2, -6)$, the midpoint of which is $(5, 0)$). That means the right triangle made between the coefficient and bullet paths at the point T has legs of length 6 and 3, so $\tan\theta = 6/3 = 2$ and we have that $x = -2$ is another root: $(-2)^3 + 14 - 6 = 0$. It works!

The last case, in the above right figure, has the point O being folded to $(2, -6)$, which means that the crease line intersects the vertical line $x = 1$ at the point $(1, -3)$. This time we have that our right triangle at O has a negative length for one of its sides; one leg has length 1 and the other has length -3. We get that $\tan\theta = -3$, and so $x = 3$ should be a root. Since $27 - 21 - 6 = 0$, this is indeed the case.

The roots of $x^3 - 7x - 6$ are so nice that they make this polynomial a perfect example for students to try Beloch's origami application of Lill's Method for themselves. In fact, this same cubic polynomial was used as the main example in Riaz's 1962 paper [Riaz62] describing Lill's Method (but not Beloch's origami version).

Follow-up

Both Lill's Method and Beloch's origami application are rich in follow-up activities.

For Lill's Method, we have already mentioned how proving Lill's Method for different cases where the coefficients have different signs (or are zero) makes for a straightforward assignment that tests if the method is being absorbed. And modeling Lill's Method in Geogebra or other dynamic geometry software is a great reinforcing, fun activity.

But there is more to be mined here. Notice how if we start with an nth-degree polynomial, then the coefficient path will have $n + 1$ sides (some of which might have length zero). Then the bullet path will have n sides and follows the same rules as a coefficient path (having $90°$ turns). That is, we can think of the bullet path as corresponding to an $(n - 1)$th-degree polynomial, and this polynomial should be equivalent to the original one but with one of the real roots factored out of it. In other words, Lill's Method is a geometric equivalent of the algebraic factoring process! (This is also explained in [Riaz62].)

As an example, look again at the $x^3 - 7x - 6$ polynomial and our Lill's Method solution that produced the bullet path for the root $x = -1$ (as seen in the figure on p. 91). This bullet path is at $45°$ angles to the x-axis, and so the lengths of the three sides of the bullet path can easily be computed to be, in order, $\sqrt{2}$, $\sqrt{2}$, and $6\sqrt{2}$. If we rotate this bullet path so that the first segment from O travels along the positive x-axis, then we get that this bullet path is the coefficient path of the polynomial

$$\sqrt{2}x^2 - \sqrt{2}x - 6\sqrt{2}.$$

If we divide this polynomial by $\sqrt{2}$, which does not affect the roots and is equivalent to normalizing the bullet path to make the first side be of length 1, we get the polynomial $x^2 - x - 6$. This makes perfect sense, since we had found the root $x = -1$ and

$$x^3 - 7x - 6 = (x + 1)(x^2 - x - 6).$$

Lill's Method really is like factoring a polynomial!

To follow up with Beloch's origami application of Lill, any student who has done Activity 7 (Can Origami Trisect an Angle?) or the construction of $\sqrt[3]{2}$ follow-up in Activity 7 will want to revisit those constructions and examine closely the "two points to two lines" fold that they employ. One will notice that neither is similar to Lill's Method. In the angle trisection construction, the lines L_1 and L_2 are not perpendicular, so it is not a secret application of Lill's Method. For the $\sqrt[3]{2}$ construction, L_1 and L_2 are perpendicular, but it is still not employing Lill's Method. (Why is left to the reader.)

This means that once students have absorbed Beloch's origami method for solving cubics à la Lill, they could create *their own constructions* for trisecting an angle or doubling a cube. For the latter, all one needs to do is to apply Lill and Beloch to the equation $x^3 - 2$. (Try it!)

For an angle trisection, one needs to use the cubic generated by the triple-angle formula $\cos 3\theta = 4\cos^3 \theta - 3\cos \theta$. If 3θ is given, then we can think of $\cos 3\theta$ as a constant k and we want to find $x = \cos \theta$. That means trying to solve the cubic polynomial

$$4x^3 - 3x - k = 0.$$

Approaching this via Beloch and Lill is easier if we fix the original angle 3θ, say $3\theta = 60°$.

Solving either of these follow-up activities requires finding a good coordinate system for your paper and making the proper setup folds. It's a very involved and hands-on process that can be loads of fun for anyone who enjoys origami geometric constructions.

Activity 10
FOLDING STRIPS INTO KNOTS

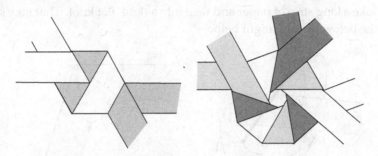

For courses: geometry, number theory, abstract algebra

Summary

Students are presented with the challenge of taking a strip of paper and folding it into a knot. (Quite simply, tying the paper into a knot in the same way one would string.) A regular pentagon should result, and the first challenge is to prove that this is indeed a regular pentagon.

Then students are asked what other knots could they make? Is it possible to make a hexagonal or heptagonal knot? What about a square or a triangle knot? What if we allow more than one strip of paper to be used?

Content

Proving that the pentagon is regular can use straightforward geometry or symmetry arguments, but the other knot explorations involve number theory and algebra. Determining what knots are possible can be rephrased as a question about the Euler ϕ function or about generators of the cyclic group $\mathbb{Z}_n$. Possibilities of using multiple strips, it turns out, is determined by the cosets of a given subgroup of $\mathbb{Z}_n$.

Handout

There is only one handout with two pages. The pages may need to be handed out separately, as the second page can give students hints for some of the questions on the first page. Use your own discretion.

Time commitment

The first page will not take much time, maybe 15–20 minutes. The second page can take longer, both because the math is more involved and because folding larger knots is a lot harder. Plan on 30–40 minutes for the second page.

Knotting a Strip of Paper

Activity: Take a long strip of paper and tie it into a tight, flat knot. That may sound weird, so the below picture might help.

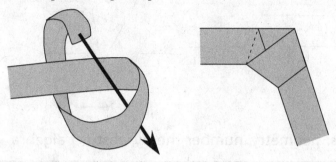

Question 1: Prove that this pentagon is regular (all sides have the same length).

Tip: When bouncing a billiard ball off a wall, the "angle of incidence" equals the "angle of reflection." Is anything like that going on here?

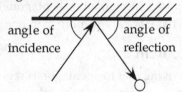

angle of   angle of
incidence   reflection

Question 2: Can you tie a strip of paper into any other knots? Hexagon, heptagon (7 sides), or octagon? How about triangle or square? Explore this and make a conjecture about what you think is going on.

Question 3: In the previous question, you should have been able to make some other knots. For example, it is possible to make an octagon knot in a number of different ways. Below is shown one way, finished off in two different fashions.

Think of each side of the octagon as being a number, starting with 0 as the side the strip entered. Then the strip weaves around and then either exits once the polygon is finished or when you get back to 0.

In what order does the paper hit the sides? Does this remind you of anything about the cyclic group $\mathbb{Z}_8$ (the integers mod 8)? Use this concept to prove the conjecture that you made in Question 2.

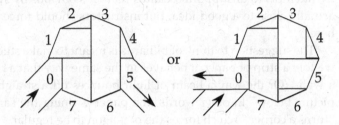

Question 4: What if we allowed ourselves to use more than one strip of paper? It turns out that then we can make just about any knot. Below are shown ways a hexagonal knot and a nonagonal (9 sides) knot can be made from 2 and 3 strips, respectively. How can the group $\mathbb{Z}_n$ be used to analyze what these knots are doing? What do the individual strips represent?

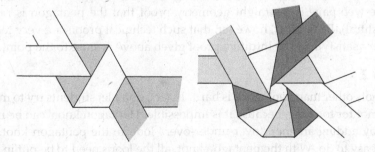

SOLUTION AND PEDAGOGY

The most important thing to have for this activity is lots of strips of paper. Those who like to teach in style can acquire quilling strips, which make very colorful knots. The more budget-minded can find large rolls of paper for accounting calculators and ticker tape in most stationary stores.

Question 1

Proofs that the pentagonal knot is regular may vary wildly, from straight geometry attempts to unsupported claims that "it's obvious by symmetry." The latter is actually close to a good idea, but instructors should force students to be specific here.

The suggestion to think of billiards is meant to make students realize that when we fold a strip of paper, it behaves in the same way that a billiard ball bounces off a wall. (Or the way a beam of light bounces off a straight mirror.) See the left picture below. In other words, the paper is doing the same thing every time it "turns a corner," which forces the pentagon to be regular.

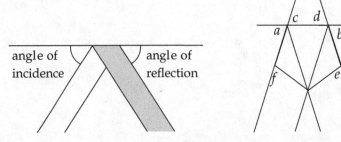

For a more rigorous way of stating this, see the right picture above. Angles a and b are the angles of incidence and reflection of one of the pentagon knot folds, so $a = b$. Angles c and d are vertical angles of a and b, so $a = b = c = d$. But c and f are angles of incidence and reflection of another fold in the knot, so $c = f$. Ditto for d and e, so we have $a = b = c = d = e = f$. Continuing in this way, we get that all the external angles of the pentagon are the same, implying that the internal angles are all the same as well. Thus the pentagon is regular.

The web site http://www.cut-the-knot.org has this pentagonal knot as its logo, and on the web page is a straight geometry proof that the pentagon is regular (.../proof.shtml). Be warned, however, that such technical proofs are very tedious and not very satisfying. The intuitive proof given above is more to the point.

Question 2

Folding knots other than pentagons is hard. In fact, don't let students try to make a hexagon knot for too long, because it is impossible. Heptagon knots can be made, however, by adding another "over-under-over" loop to the pentagon knot. No, this is not easy to do. With the pentagon knot, all the loops need to be put in place and then the knot is slowly tightened. The same thing needs to happen in the heptagon case, only with more loops involved that are much more able to slide around and cause trouble. Use the below figure as a guide for how to keep things arranged before the knot is tightened.

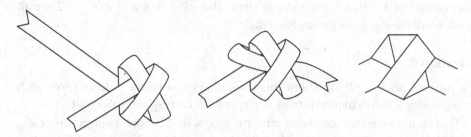

Making such knots takes patience and practice. After struggling with your first heptagon knot, make another one—it will be a lot easier and come out much better. Students won't be too surprised at the difficulty of this, but you should try to make a few yourself to show them.

The conjecture students should be striving for is that any regular polygon can be knotted in this way from a single strip of paper *except* for the triangle, square, and hexagon. Experimentation should allow students to arrive at this result, and they should be encouraged to try proving it before turning to the second page of the handout, which provides hints.

In fact, instructors may want to hand out the two pages of the handout separately. First of all, the first page, as written, could be used in any number of courses, starting as "low level" as a math for liberal arts class. Only the second page uses language of group or number theory. Secondly, if students turn to the second page too quickly, they will get hints as to what kinds of knots can be made. That's not necessarily a bad thing. But the two pages don't have to be handed out at the same time.

Question 3

The advantage of numbering the sides of the knotted n-gon with the integers $0, 1, \ldots, n-1$ is that then we can think of folding the knot as following a suitable cycle in the group $\mathbb{Z}_n$.

Specifically, if we begin our strip at 0, then we want the strip to travel across the n-gon and come out on some side numbered $2, \ldots, n-2$. Suppose that it comes out on side a. Then the strip will turn a corner, and by the angle of incidence/reflection argument it will next come out at the $2a$ side. (It must leave side a at the same angle by which it entered, and this preserves the number of sides we "skip" as the strip bounces around the polygon.) So we need to hit every side of the polygon for the knot fold to work, and this will happen if and only if a generates the whole group $\mathbb{Z}_n$.

To sum up: We will be able to fold an n-gon if and only if $\mathbb{Z}_n$ has a group generator that is not 1 or $n-1$. In other words, if and only if there exists an element of $\mathbb{Z}_n$ other than 1 or $n-1$ that is relatively prime to n. In other words, if and only if $\phi(n) > 2$, where ϕ is the Euler ϕ function, the number of positive integers $< n$ that are relatively prime to n (including 1).

We have that $\phi(3) = \phi(4) = \phi(6) = 2$, so polygons with those numbers of sides cannot be knotted from a single strip. But $\phi(5) = 4$ and $\phi(n) > 2$ for all $n > 6$, so all other n-gons can be folded.

Question 4

The fun and amazing thing about this question is that when making knots with multiple strips, the individual strips of paper each correspond to a coset.

That is, to make an n-gon with multiple strips, first choose a subgroup of $\mathbb{Z}_n$; call it H. If $|H| = k$, then there will be n/k cosets of H (including H itself), and this means that we'll need n/k strips of paper.

For example, in the nonagon picture on the handout, I chose the subgroup $H = \{0, 3, 6\}$ in $\mathbb{Z}_9$. This can be thought of as, say, the white strip of paper, which starts on the 0 side, goes to side 3, bounces to side 6, and returns to side 0. Then another strip is needed to cover the sides $1 + H = \{1, 4, 7\}$ (the dark grey strip) and another to cover the sides $2 + H = \{2, 5, 8\}$ (the light grey strip). This covers the whole group, so those three strips will complete the 9-gon.

Such multiple-strip knots are *not* easy to fold, but such a hands-on and direct application of cosets is too much to resist for an algebra class. It even demonstrates Lagrange's Theorem. Also, multiple-strip knots can be made to weave in very symmetric patterns (as done in the nonagon picture) and thus result in very attractive rings when made from strips of different colors.

If you are brave enough to attempt one of these multiple strip knots (other than the hexagon, which is easy), I recommend making a 12-gon knot out of three strips of paper. While difficult, this one is a bit easier because each strip will follow a coset of $\{0, 3, 6, 9\}$ which will form a square. Thus the folds will all be at $45°$ angles to the side of the paper. The length of paper between these folds still needs to be determined by trial and error (or you could challenge your students to determine the exact length needed and then measure it with a ruler), but with experimentation and tweaking this can make a very attractive woven ring.

Background

People seem to have known about folding pentagonal knots in strips of paper for a long time. According to Fukagawa, it forms the subject of a Japanese *sangaku* dating back to 1810 [Fuj82]. *Sangaku* were geometry problems artfully written on wooden tablets and hung in Shinto temples during Edo-era Japan (1600s–1800s). They form excellent evidence that common people in ancient Japan would play with recreational geometry problems, and the fact that some of these *sangaku* were about paper folding means that some Japanese of that time were interested in the mathematics of origami. (See the Haga's "Origamics" activity for another example of an origami *sangaku*.) This particular *sangaku* depicts a picture of a pentagonal knot folded from a strip of paper and asks the reader to determine the relationship between the width of the paper strip and the side length of the pentagon.

References to larger polygonal paper knots are more rare. One of the earliest seems to be by Morley in 1924 [Mor24], who shows instructions for pentagon, hexagon, and heptagon knots and generalizes them.

Activity 11
HAGA'S "ORIGAMICS"

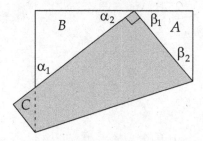

For courses: geometry, math for liberal arts, introduction to proof

Summary

Kazuo Haga's "Origamic" activities ask students to explore simple, geometric properties found when we fold paper in a prescribed way. The aim of these activities is to give students an easy-to-explore paper folding puzzle so that they can experience a micro-version of the three stages of mathematical research: exploration, conjecture, and proof.

Content

Haga's activities are all geometry-based. Some require no prior knowledge at all, while others make use of some standard Euclidian geometry facts. The methods of discovery, conjecture, and proof, however, lie at the heart of all these exercises.

Haga's activities have been published, in Japanese, in a book [Haga99], in several articles in the now-defunct Japanese origami magazine *ORU* [Haga95], and in English in the book [Haga08]. Most of the material presented here, however, is reproduced in Haga's article "Fold Paper and Enjoy Math: Origamics" in the proceedings book *Origami³: Third International Meeting of Origami Science, Mathematics, and Education* [Haga02].

Handouts

There are four handouts, each an example of Haga's origamics. Each activity will require lots of small squares of paper for each student (three-inch memo cube paper is ideal).

(1) Folding TUPs.

(2) All Four Corners to a Point.

(3) Haga's Theorem.

(4) Mother and Baby Lines.

Time commitment

Each activity will take a whole 50 minute class, although it does depend on the level of your students.

Folding TUPs

Take a square piece of paper and label the lower right-hand corner A. Pick a random point on the paper and fold A to that point. This creates a flap of paper, called the Turned-Up Part (or TUP for short).

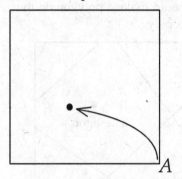
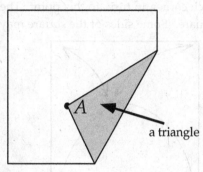

a triangle

How many sides does your TUP have? Three? Four? Five?

Your task: Experiment with many TUPs to find an answer to the question, "How can we tell how many sides a TUP will have?"

Follow-up: What if we allowed the point to be outside the square? Then what are the possibilities?

Haga's Origamics: All Four Corners to a Point

Take a square piece of paper and pick a point on it at random. Fold and unfold each corner, in turn, to this point. The crease lines should make a polygon on the square. (Some sides of the square may be sides of this polygon.)

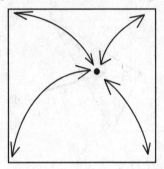
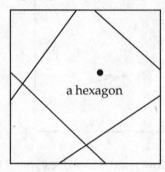

a hexagon

How many sides does your polygon have? Five? Six? Could it have three, four, or seven?

Your task: Do this "all four corners to a point" exercise on many squares of paper. How can you tell how many sides your polygon will have?

Follow-up: What if we used a rectangle instead of a square? Then what are the possibilities?

Haga's Origamics: Haga's Theorem

Take a square piece of paper and mark a point P at random along the top edge of the paper. Then fold the lower right corner to this point.

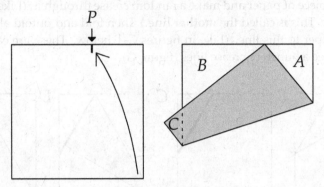

Question 1: What nice relationship must be true about the triangles A, B, and C? Proof? (This is known as Haga's Theorem.)

Question 2: Suppose that you took the point P to be the midpoint of the top edge. Use Haga's Theorem to find out what the lengths x and y must be in the below figure.

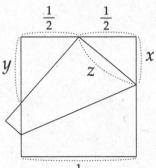

Haga's Origamics: Mother and Baby Lines

Take a square piece of paper and make a random crease through it. (Like in figures A and B below. This is called the **mother line**.) Then fold and unfold all the other sides of the paper to this line. (Like in figures C–F below. These are called **baby lines**.) You'll see a bunch of crease lines (figure G).

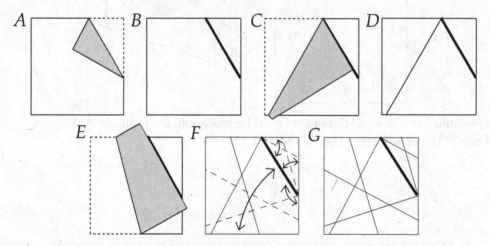

Your task: Experiment with various mother lines on separate sheets of paper and compare your results. What conjectures can you make about the intersections of the baby lines? Prove it/them.

SOLUTION AND PEDAGOGY

Kazuo Haga is a retired professor of biology from the University of Tsukuba, Japan. Since his retirement, he has been running Haga's Laboratory for Science Education, where he promotes his "origamic" activities as a way to develop scientific reasoning skills among children and students. Haga invented the term *origamics* as a way to describe the scientific side of paper folding, since often origami is thought of (especially in Japan) as only an activity for children.

Do note, however, that Haga's activities pose some serious challenges for both teacher and student. They are deliberately open-ended, so that students will be forced to experiment, make conjectures, and then try to prove them. At the same time, some instructors may be faced with students, even math majors in an introduction to proofs course, who are resistant to such open-ended assignments. Keeping such students motivated and on task might be difficult, and instructors will need to figure out what works for their kind of students. Perhaps grades or the "glory" of getting a conjecture or theorem named after themselves (for internal class use) will be enough motivation. In any case, these activities are asking students to think in sophisticated ways, and instructors should not underestimate how difficult this can be.

Handout 1: Folding TUPs

Students should quickly realize that as long as the random point is chosen to be inside the square, only triangle and quadrilateral TUPs are possible.

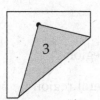 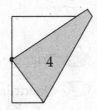 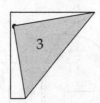 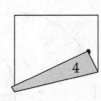

Experimenting shows that when we pick the point nearby the main diagonal of the square—the one going from the point A to the opposite corner—we always get a triangle. A more careful look shows that a quadrilateral will result only if a whole side of the paper is folded over. Thus we can think of the two sides of the square that A lies on as acting as radii of circles (with A on the circumference); if A is folded to a point beyond one of these radii then the corresponding corner (either the upper right or the lower left) will be folded over, creating a quadrilateral. Thus, we can color the square into the regions shown below, thus solving the problem. (Note that the boundaries belong to the triangle region.)

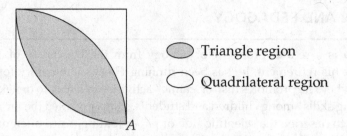

Follow-up. When allowing ourselves to fold the corner A to points outside the square, we need to think about when this would and when it wouldn't make sense. Clearly if we fold A to a point very far away from the square, the "fold" would be tantamount to just flipping the entire square over. But if we fold A to a nearby outside point, we can get pentagonal regions for our TUP.

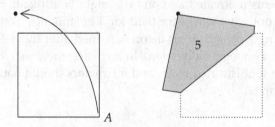

This creates a "new radius" to consider. Or rather, we get a pentagon if *both* of the sides adjacent to A get folded over. This creates a circle centered at the corner opposite A whose radius is the length of the diagonal of the square. (See below.)

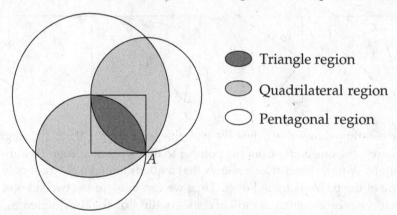

While the solution above is perfectly fine and rigorous, more advanced students might devise other methods of proof for this problem. Andy Miller (Belmont University) and his students came up with an analytical approach. Suppose that the square is in the xy-plane, with the lower left corner at the origin and A at the point $(1,0)$, and that we fold A to the point $P = (a,b)$. The crease line will be the perpendicular bisector of the segment $\overline{AP}$. We can find the equation of this crease line using the same methods as in the Folding a Parabola activity. The slope of $\overline{AP}$ is $-b/(1-a)$, so the slope of our crease line is $(1-a)/b$. The midpoint of $\overline{AP}$, $((a+1)/2, b/2)$, is on the crease line. Thus, the crease equation is $y - b/2 = ((1-a)/b)(x - (a+1)/2)$, which simplifies to

$$ y = \frac{1-a}{b}x + \frac{a^2 + b^2 - 1}{2b}. $$

The y-intercept of this line is $(a^2 + b^2 - 1)/2b$, and if the lower left corner is going to fold over (and thus make the TUP have more than three sides), then this y-intercept has to be greater than zero. That is, we'd want $a^2 + b^2 - 1 > 0$, or $a^2 + b^2 > 1$. This describes one of the circular regions. Equations for the others can be obtained by similar means.

Pedagogy. The handouts for Haga's origamics are deliberately open-ended. The point is for students to come up *themselves* with the idea of shading different regions of the paper to indicate different TUPs and to generate enough data to get a feel for the proper picture. That's the experiment-conjecture-proof method that lies at the heart of mathematical research.

Still, there are many leading questions/suggestions an instructor can give to help students along, although there is a lot of value to letting groups of students hammer away at this activity for extended periods of time. The following list of tips could be given to guide students in their thinking:

(1) Experiment with many choices for the target point P to get data, from which you might make a conjecture. You could even be systematic about it by forming a grid of points, letting P be each of these points, and coloring the point depending on how many sides its TUP had.

(2) Following this lead, try to think of what region of the square yields choices for P that give triangle TUPs and what regions give other TUPs.

(3) After thinking about that, can you try to nail down where the *boundaries* between these regions are? For example, if you move the point P around, when will it change from a triangle TUP to a quadrilateral TUP?

Handout 2: All Four Corners to a Point

Students may think that only pentagon and hexagon regions can be made, but there are a finite number of places where a quadrilateral region is formed: the exact center and the four corners. Those are more like anomalies, however. The only regions with nonzero area are those that create pentagons and hexagons.

Actually, this problem is very similar to the TUP handout. With TUPs, we needed to keep track of whether the chosen point causes a corner of the paper to turn over. With the current activity, we care about whether or not *midpoints of the sides* get folded over. See the figures below to see why. If the point P to which we're folding is far enough away from a side of the square, then the two crease lines made by the corners of that side will intersect at a point on the square. On the other hand, if P is close enough to the side, then the two crease lines will intersect at a point not on the square. This will determine whether or not those two crease lines will add two or three sides to the region containing P on the square.

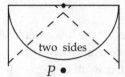   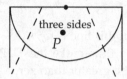

The difference between being "far enough away" and "close enough" is determined by drawing a circle of radius 1/2 the side of the square centered at the midpoint of the side. (If a corner adjacent to the side gets folded outside of this circle, then the midpoint moves when the corners are folded. Otherwise it stays put.)

Then, we need to see how all four of the sides will interact. The four circles centered at the midpoints will only intersect in pairs. Now, if we momentarily ignored the sides of the square, then for a given point P, the region containing P determined by the crease lines will always be a quadrilateral. But then the sides of the square will cut off a number of corners of this quadrilateral equal to the number of circles P is inside. The most number of circles P could simultaneously be inside is two, generating a hexagon. Otherwise P will be in only one circle, giving a pentagon (excluding the five cases mentioned earlier where P is not in the interior of any circle). Overlapping the four circles provides a nice picture of the hexagon and pentagon regions, as shown below left.

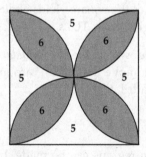   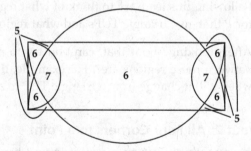

What about if P lies on the boundary of one of the circles? Being on the boundary means that the two creases intersect on the boundary of the square, so a new side of the point's region will not be generated. So boundary points will be part of the pentagon regions.

Follow-up. The same analysis works with rectangle paper, but the surprising thing is that heptagon regions are now possible! (See the right figure above.)

Pedagogy. As with the TUP activity, the whole point of this is for students to analyze and create a model of the situation on their own. So very few hints should be given by the instructor, aside from clarifying the nature of the problem.

In my experience, students who work through the TUP activity pick up on this one pretty quickly. In fact, this one could be given for homework if the TUP one is done in class.

Handout 3: Haga's Theorem

This activity might be the first such "origamic" result that Haga developed, which may be why people in the Japanese origami community named it after him. But since then an example of a Japanese *sangaku* (geometry problems written in 1600s–1800s Japan and left in religious temples for other people to read and solve) was found implying that this result was known to Edo-era Japanese geometers. (See [Fuk89] p. 37 and p. 117.) Nonetheless, it seems Haga was unaware of this obscure reference.

Question 1. The basic result is that the triangles A, B, and C on the handout are *all similar*. The proof is simple. In the figure below we have that $\alpha_2 + \beta_1 = 90°$ (since $\alpha_2 + 90° + \beta_1 = 180°$), and we also know that $\beta_1 + \beta_2 = 90°$. So $\alpha_2 = \beta_2$, and similarly $\alpha_1 = \beta_1$. Thus $A \sim B$, and the same reasoning shows that $B \sim C$ as well.

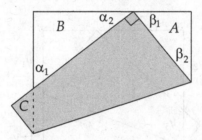

What has made Haga's Theorem so popular among origamists is the fact that this simple one-fold move creates a figure with a wealth of elegant geometrical aspects. The main application for origami purposes is that Haga's Theorem can give us simple solutions to the problem of dividing the side of a square into $1/n$ths, where n is some odd number. An example is seen in the next question.

Question 2. There are many tools at our disposal with which to find the lengths of the sides x, y, and z in the figure. Namely, we have the similar right triangles to work with, as well as the fact that $z = 1 - x$, since the segment of length z is the image of the $1 - x$ segment under our fold. So, using the Pythagorean Theorem on triangle A,

$$1/4 + x^2 = (1 - x)^2 \implies x^2 = 3/4 - 2x + x^2 \implies x = 3/8.$$

Notice that since the x^2 terms canceled out, we found that x is a rational number. Thus $z = 5/8$ is also rational. Now, to find y, or just about any other length of the segments that Haga's Theorem generates, we'll take advantage of the similar triangles. Since this involves only comparing ratios, we know that y will also be a rational number too!

This means that Haga's Theorem can be applied to obtain rational divisions of the side of a square, and if we're lucky then these rational divisions might turn out to be useful. Indeed, $A \sim B$ gives

$$2y = \frac{1}{2x} \Rightarrow y = \frac{2}{3}.$$

Of course, the lengths of all the segments in the handout figure can be found in this way. In fact, to see the full power of Haga's Theorem, let the placement of the point P be arbitrary and then compute the lengths. If we label these lengths as in the figure below, letting the length to the right of P be x, then the remaining lengths become

$$y_1 = \frac{(1+x)(1-x)}{2}, \; y_2 = \frac{2x}{1+x}, \; y_3 = \frac{1+x^2}{1+x}, \; y_4 = \frac{(1-x)^2}{2}$$

$$y_5 = 1 - \left(\frac{2x}{1+x} + \frac{(1-x)^2}{2} \right), \; y_6 = \sqrt{x^2+1}.$$

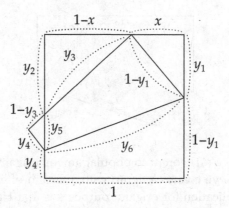

Pedagogy. In a sense, developing and exploring Haga's Theorem is merely an intense application of similar triangles, the Pythagorean Theorem, and basic algebra. One could envision it as a great activity for a precalculus or other basic algebra class, except the motivation tends to get lost on such students. In fact, any student other than a math major would likely not be impressed by the elegant, one-fold manner in which Haga's Theorem gives us a bevy of rational lengths. This is why in the handout I chose to include only one example of this, showing how the length 1/3 can pop out. This can be motivated somewhat, since no student will likely be able to know of any other means by which a square piece of paper could be divided into perfect thirds (unless they've done the Dividing a Length into $n$ths Exactly activity).

Still, everything about this activity is doable by any college student. The concept of similar triangles is sometimes given short shrift in high-school geometry classes, but it, along with the Pythagorean Theorem and basic algebra, are things anyone from general education students on up can be expected to know.

Thus when groups of students are working on this handout, very few hints should be given by the instructor. Hinting that the triangles should be examined could be offered, or possibly posing Socratically, "What do we remember about right triangles?" but no more.

Haga's Theorem can be modeled, to great effect, on Geogebra or Geometer's Sketchpad. One would have to construct a square and then do something similar to the Folding a Parabola activity to construct the crease line made when folding the lower right corner to a random point P on the top edge. (Construct the segment connecting these points, and then the crease will be the perpendicular bisector.) Then use Geogebra to reflect the bottom part of the paper about this crease line. The advantage of doing this is then you can have Geogebra measure the lengths of the various line segments and then students can see how they change as you move P back and forth along the top edge. In this way, it becomes very easy to see what you get if P is at the 1/4 mark, or the 1/3 mark, or the 2/3 mark, etc.

Of course, there's much more that can be done with Haga's Theorem. In his book *Geometric Origami* [Ger08], Austrian geometry teacher Robert Geretschläger proposes a series of interesting facts about Haga's Theorem. Namely, if we consider the labeling as shown below, suppose that we draw a circle centered at the square's corner C with radius equal to a side of the square. Then this circle will be tangent to the line $C'D'$. This can then be used to prove that the perimeter of $\triangle AGC'$ is equal to half the perimeter of the original square, and that the sum of the perimeters of triangles $C'BE$ and $GD'F$ is equal to the perimeter of triangle AGC'. (This problem was posed in the 37th Slovenian Mathematical Olympiad, 1993.)

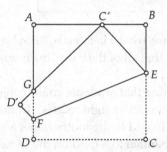

Additionally, one could ask if Haga's Theorem could be generalized, as suggested above, by considering different locations for the point P. If you or your students pursue this, you'll find that we could have included Haga's Theorem in the Dividing a Length into $n$ths Exactly activity. One can prove that any odd division of the side of the square can be found by a variation of of Haga's Theorem. See

Haga's book *Origamics: Mathematical Explorations Through Paper Folding* [Haga08] for more information.

As you can see, Haga's Theorem is rich with secrets.

Handout 4: Mother and Baby Lines

Of the four origamics exercises presented here, this one is the most challenging. Developing the proper conjectures will take some experimentation and creativity, as they are not obvious.

As students (or professors too) struggle to find patterns in the intersecting baby lines, it can be helpful to remember that half the battle when researching a wide-open problem is to *ask the right questions*. For example,

- Do the baby lines seem to intersect at any interesting angles?

- Is there any significance to the number of baby line intersections on each side of the mother line?

- Does anything interesting happen when we choose the mother line to be something symmetric, like a diagonal of the square or a "fold in half" vertical line?

- Are any three of the intersection points collinear (other than obvious cases)?

Exploring questions like these can uncover many things that are going on here. The figure below tells part of the story (at least in one example).

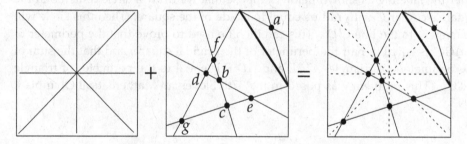

In the left-most picture are what Haga calls (see [Haga02]) the *primary crease lines of the square*. We can see that all of the baby line intersection points seem to lie on these primary lines!

But that's not the only thing that students could conjecture. Notice that some of the baby lines seem to intersect at right angles. In fact, if we consider a baby line made by a side, call it *S* folded to the mother line, and then consider a second baby line made from a side parallel to *S* (and on the same side of the mother line), then these two babies seem to intersect at right angles. Do they?

These conjectures can be proven using a few applications of Euclidean geometry. The most basic example of this can be seen in the intersection point labeled *a* above. This point lies inside a right triangle made by the mother line and the top and right sides of the square. In fact, the baby lines made in this triangle are angle bisectors of the triangle. So the point *a* is actually the *incenter* of this triangle, and thus it will also lie on the angle bisector of the other corner of the triangle, which just happens to be a primary line of the square. Bingo!

The other intersection points can be explained if we extend the mother line as well as some sides of the square. Point b, for example, is also the incenter of the right triangle made from the extended mother line and the extended bottom and left sides of the square, which proves that it lies on a diagonal. This point is crucial; we can see that when we fold a side segment S of the square to the mother line, if S is adjacent to the mother line, then the crease made will be a bisector of the angle made by S and the mother line. But if S is not adjacent to the mother line, then the crease made *is still an angle bisector*, but now it's bisecting the angle made by the mother line and S extended so they intersect off the piece of paper.

Now consider the point c, where we can extend the mother line and the left side of the square to view the angle that one of c's creases is bisecting, as shown below, forming $\triangle ABC$ with the other baby crease line at c. We have two things to prove: (1) That c also lies on a primary crease line and (2) that the baby lines meeting at c intersect at right angles.

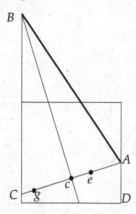

It can be very easy for students to accidentally assume what they're trying to prove here. For example, students may claim that the point C lands on the point A when they fold crease line Bc, which proves the babies are at right angles. But the baby line AC was made by folding the segment AD to the mother line, and that does not directly imply that C must land on A when we fold Bc. A student's folded example may provide good evidence that this is indeed happening, but that's not a proof! We must be wary of "proofs by origami" since we can't always trust everything we see happening on folded paper.

A better approach is to note that since the left and right sides of the square are parallel, we have that the alternate interior angles they make with AC are equal. That is, $\angle BCA = \angle CAD$. But by the definition of AC, we know that AC bisects $\angle BAD$, so $\angle CAD = \angle CAB$. Thus $\angle CAB = \angle BCA$, which proves that $\triangle ABC$ is isosceles.

This immediately gives us what we want. The base of an isosceles triangle is perpendicular to the angle bisector opposite it, which gives us the perpendicular baby lines. And since segments Cc and Ac are congruent, the point c must lie on the vertical half-way line of the square, which is a primary line. A similar argument will further show the same results for the baby line intersection point d.

For points e and f, we can see that they lie on the *excenter* of some well-chosen triangles. An excenter of a triangle is the intersection of one of the interior angle bisectors and two exterior angle bisectors. It is the center of a circle, outside the triangle, that is tangent to one side of the triangle and extensions of the other two sides. This is more easily seen in the left figure below, where we consider $\triangle BEF$. (This is just the top part of the isosceles triangle that we considered previously.) We see that the baby line intersection point f lies at the intersection of the internal angle bisector at B and the external angle bisector at F. Thus f lies at the excenter off the bottom side of $\triangle BEF$, and thus the external angle bisector at E will also pass through f, which happens to be a primary crease line. A similar argument can be used for the point e.

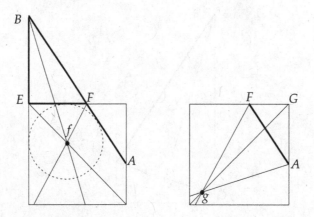

Now we are left with the point g. According to Haga himself, "In mathematical classes or courses, students find the most difficulty in proving that this point is on a primary crease." [Haga02] However, g is also the excenter of a triangle. As seen in the right figure above, the baby lines that define g are external angle bisectors of $\triangle GAF$, and the interior angle bisector for this excenter is a primary crease line.

One need not rely on excenters to get these results, however. Geometry students will likely find other proofs. As an example, Haga offers the following alternate proof for the point g: Draw perpendicular lines from g to the right side of the square (gH), the top side (gJ), and the mother line (gI), as in the next figure. Then, since the baby lines (Fg and Ag) are angle bisectors, we get that $\triangle gAH \cong \triangle gAI$ and $\triangle gFI \cong \triangle gFJ$. Thus gJ, gI, and gH all have the same lengths. That is, the point g is equidistant from the top and the right sides of the square, meaning that it must lie on a diagonal of the square.

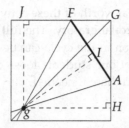

The advantage to using proofs based on incenters and excenters, however, is that they generalize easily. Our proofs above were only for one specific case, but the principles at play with in- and excenters will work with any choice of mother line. Generalizing proofs that use dropping perpendiculars and such may not be so easy.

There are many other explorations that can be made with this activity. For example, notice that when the mother line is taken at random, the babies on either side of it form a collection of lines in general position. Thus the number of baby line intersection points on either side of the mother line will, in general, be a triangular number. This assumes, of course, that all the intersection points lie on the square. In fact, if we take into consideration *all* baby line intersections, even the ones that occur outside the square, we get similar results if we allow the primary crease lines to tessellate in a square grid determined by the square paper.

A more ambitious follow-up project would be to explore similar results for arbitrary, convex polygonal paper. In such cases, the primary crease lines might be taken to be the angle bisectors of the corners of the paper...

Pedagogy. The main conjecture to be made in this activity, that all the baby line intersection points lie on primary crease lines, is by no means obvious. There's a good chance that a classroom-full of students will not stumble upon this observation. Instructors can increase this probability, however, by making sure that the students do their explorations carefully. It's a very good idea that students draw their mother line (after they've folded it) with a pen for emphasis. Then as the baby lines are made, suggest that they draw a dot at their intersection points. If anything, this will make their work neat.

But it's important that they try a number of examples with different mother line choices. In fact, it can be interesting (and perhaps suggestive as to what's going on) to let the mother line be one of the primary lines itself, as suggested in the questions stated previously. For example, if the mother line is a diagonal of the square, then we only get two baby line intersection points, both of which lie on the other diagonal. Looking at simple cases like this, where there's not much to observe except for the "obvious" fact that the points lie on the other diagonal, may inspire students to look for similar behavior on more complex examples.

If anything, the origamics examples provided here illustrate the amazing variety and depth that very simple origami geometry exercises can display. Haga has many more such activities, and people should try inventing their own. But aside from being just fun, the real worth of these activities is how each one offers us a mathematical research micro-laboratory. The full gamut of exploration, conjecture, proof, disproof, and so on in the cycle can be found in each of them. As teachers, all we need to do is let our students loose on them, then sit back and watch, offering a nudge here and there if needed.

Activity 12
MODULAR STAR RING

For courses: precalculus, geometry, math for liberal arts

Summary

Instructions are given to make a modular origami star-shaped ring. Strangely, the instructions do not say how many units need to be made in order to make the ring! Students are then asked to calculate the ideal, or perfect, number of units needed to make this ring.

Content

The units for this ring are very easy to fold, and so students should have no trouble folding lots of units quickly. The ring can be made using any number from 12 to 16 units, and question is, then, what is the "correct" number of units to make them fit together perfectly into a ring. Doing this requires a careful analysis of the unit, simple triangle geometry, and either a careful observation of the angles or the formula for the sum of the angles in a polygon. As a follow-up, the radius for the completed star can be calculated, again using geometry.

Handout

The handout contains instructions for the modular ring and asks the basic question of figuring out how many units are needed. An optional second page is also included to offer students a hint on how to approach the problem.

Time commitment

This model does require a good number of units, at least 12 per student. So the folding part of this activity will take about 20–30 minutes. Answering the question will vary in time depending on the type of class being taught, but if the optional "hint" page is provided, expect 5–10 additional minutes to solve the problem.

Modular Star Ring

This unit makes a star-shaped ring. You will need about 12–20 squares of paper.

(1) Fold and unfold in half in both directions.

(2) Fold all four corners to the center.

(3) Fold the top edges to the center vertical line.

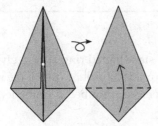 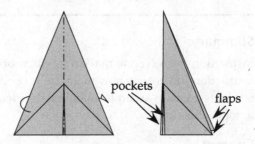

(4) Turn over and fold the bottom point up as shown.

(5) Fold in half **away from you** and you're done! Make a bunch more.

Putting them together: Slide the flaps of one unit into the pockets of another. (The pockets and flaps are indicated in step (5) above.)

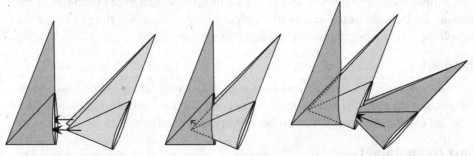

Keep adding more units until it comes back to the first unit and forms a ring!

Question: You may have noticed that you can make this ring close up with 12, 13, 14, or even more units, but some of these feel pretty loose. What is the best number of units you should use to make a tight, perfectly-fitting ring?

Additional hint: To make the units fit perfectly, you want each unit to slide in as far as it can, with the top edge of each unit's flaps "flush" against the top edge of its neighbor's pockets.

The below picture might also help you see what the proper angles should be if the units are perfectly fitted together.

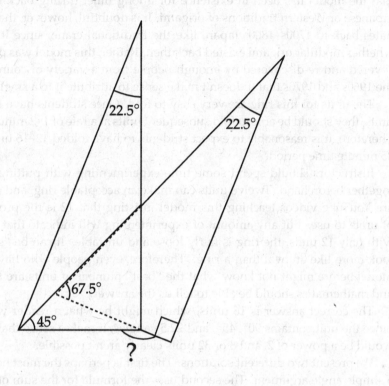

SOLUTION AND PEDAGOGY

The origins of this modular ring do not seem to be known. The web page of OrigamiUSA (the national not-for-profit origami society) contains diagrams of this model, and they list it as a "traditional" model. This phrase is intended to convey that the model has been in existence for a long time, tracing back to either the Japanese or Western traditions of origami. It is doubtful, however, that this model dates back to 1700s–1800s Japan, like the traditional crane, since it is not clear whether modular origami existed back then. Rather, this model was probably discovered and re-discovered by enough people from a variety of countries during the 1960s and 1970s that it doesn't make sense to attribute it to a single person.

The units for this ring are very easy to fold. Once students have folded a few units, they should be able to fold subsequent units at a rate of 1–2 minutes per unit. Therefore, it is reasonable to expect students to have folded 12–16 units during a 25 minute time period.

Instructors should spend some time experimenting with putting these units together beforehand. Twelve units can make an acceptable ring, and in fact, there are YouTube videos teaching this model insisting that 12 is the proper number of units to use. But any amount of experimenting will indicate that when made with only 12 units, the ring is fairly loose and unstable. It can be "smushed" to look more like an oval than a ring. Therefore, even people who have made this model before might not know what the "best" number of units are for this ring, and mathematics should be able to tell us the answer!

The correct answer is 16 units, which might be what a reader would guess. Since the unit contains 90°, 45°, and 22.5° angles, it makes sense that the answer would be a power of 2, and 8 or 32 units clearly aren't possible.

We present two different solutions. The first is perhaps the most natural, using a simple angle argument. The second uses the formula for the sum of the interior angles of a polygon (thus making this a good activity for any class that includes this formula in its curriculum).

Solution 1

The most simple solution would be to determine how much one unit rotates with respect to an adjoining unit. Since this angle of rotation would remain the same from one unit to the next, we can just divide 360° by whatever this angle is and get the number of units we would need.

Each unit is a right triangle with angles 22.5°, 67.5°, and 90°. These angles can be determined by studying the folding sequence of the units and realizing that the tip of each unit, where the smallest angle resides, is a corner of the original square that has been bisected twice, thus producing a 22.5° angle.

Next, the angles involved when two units are joined need to be determined. To do this, a good picture of two units locked together helps tremendously. This is why the optional second page for the handout is included. Below is a reproduction of this figure with points *A*, *B*, *C*, and *D* labeled as well as one other angle.

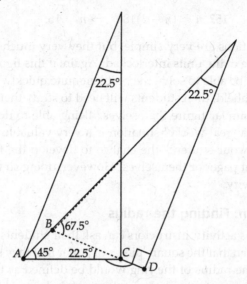

We have that $\angle ABC = 180° - 67.5° = 112.5°$, and so $\angle BCA = 180° - 45° - 112.5° = 22.5°$. In other words, to go from the unit on the right to the unit on the left we rotate by exactly 22.5°. To make a complete ring we would need to rotate a full 360°, and so the number of units we need is

$$\frac{360°}{22.5°} = 16.$$

It should be noted that in the author's experience, students have not been likely to make this simple argument. It requires the student to look at the problem from the perspective of the units themselves, whereas it seems that people more naturally look at the problem from the perspective of the whole ring. From that perspective, the next solution is the more natural one.

Solution 2

If one makes a complete ring, one sees that the center is an open polygon. If the units are inserted into each other all the way (as in the instructions and the above figure), then the interior angles of this polygon hole should all be the same. If we can calculate what this common interior angle should be, then we can calculate the number of units needed by using the equation that the sum of the interior angles of a polygon with *n* sides equals $(n - 2)180°$. Here *n* would equal the number of units.

The above figure shows that the interior angle of the polygon hole is

$$\angle ACD = 180° - 22.5° = 157.5°.$$

Therefore, if $n =$ the number of units, which equals the number of sides of our polygon hole, then we have that the sum of the interior angles is $157.5°n$ and

$$157°n = (n-2)180° \Rightarrow n = 16.$$

Both of these solutions are very simple, but they very much rely on having a carefully drawn figure of the units interlocked. Again, if this figure is given to the students, they should be able to solve the problem quite quickly.

If the figure is withheld, then students will need to study their units very carefully and produce a similar figure themselves. Being able to do this kind of geometric modeling of a "real-world" situation is a very valuable skill, and so it is encouraged to allow your students the chance to develop the figure on the optional second handout page for themselves. However, doing so will require more class time for the activity.

Further exploration: Finding the radius

As an extension to this activity, instructors can ask their students to determine the radius of the ring, given that the square pieces of paper that they started with have side length 1. Here the radius of the ring would be defined as the distance from one of the points of the star to the center of the ring's hole.

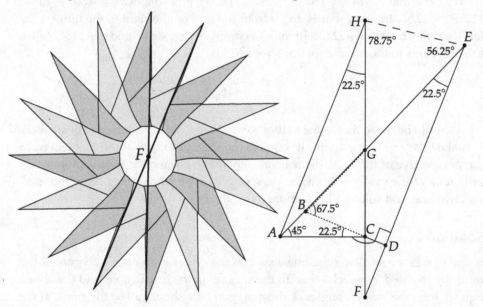

The above figure shows how the center of the ring F is related to the configuration of the units; it is at the intersection of the extended right sides of two interlocked units. We can transfer this to our previous figure to incorporate the ring's center, labeled as point F in the above-right figure.

Since we are assuming that our square pieces of paper have side length 1, studying the folding sequence will show that the right side of the units, line ED

in the above-right figure, has length $\sqrt{2}/2$. (It is the length of one of the diagonal creases shown in step (2) of the instructions.) The shorter side of the unit, which is BD in the figure, has length $(\sqrt{2}-1)\sqrt{2}/2 = (2-\sqrt{2})/2$. How did we get that? These triangles are $22.5°$-$67.5°$-$90°$ triangles, so we can use the canonical side lengths of such a triangle, as found in Activity 2 (Origami Trigonometry). We know the length of ED, so to find BD we can set up the ratios with the canonical lengths to get $BD = (2-\sqrt{2})/2$.

There are several ways to proceed from here, and all of them (to my knowledge) lead to some fairly yucky nested radicals in the final answer. My preferred method, however, is to focus on $\triangle GEH$ in the above-right figure. Notice that since $\triangle EGF$ is isosceles, we have that $EG = FG$, and so the radius of our star, $FH = FG + GH$, can be found by adding GH and EG. That is, we want to add two of the sides of $\triangle EGH$.

GH is the easier side length to find. Since $\triangle AGH$ is also isosceles, $GH = AG$, and AG is the hypothenuse of the $45°$ right triangle $\triangle ACG$. Now, $AC = BD = (2-\sqrt{2})/2$ as we saw before, so

$$GH = AG = \sqrt{2}\frac{2-\sqrt{2}}{2} = \sqrt{2}-1.$$

Length EG is *much* harder to compute. Let us use the law of sines to do it. Triangle $\triangle EFH$ is an isosceles triangle whose angle at F is $22.5°$. Thus $\angle GHE = (180° - 22.5°)/2 = 78.75°$, and therefore $\angle GEH = 180° - 78.75° - 45° = 56.25°$. (Note that $\angle EGH = 45°$.) These are very non-traditional angles, and if we hope to get an exact answer, then we will need to know the exact value of sine of these two angles.

Note, however, that middle- or high-school students could simply get decimal approximations of these lengths using a calculator (either using the law of sines or in other ways, like finding similar triangles and trying to compute the length DF). That is fine. But we will proceed in the quest for an exact solution.

First we will find $\sin 78.75°$. Since $90° = 78.75° + 11.25°$ and $11.25° = 22.5°/2$, we could use the methods form the Origami Trigonometry activity to find $\cos 11.25° = \sin 78.75°$. Or we could just use the reverse of the double-angle formula for cosine, $\cos 2x = 2\cos^2 x - 1$, which means that $\cos x = \sqrt{(\cos 2x + 1)/2}$. We found $\cos 22.5° = \sqrt{2+\sqrt{2}}/2$ from the Origami Trigonometry activity, and so

$$\sin 78.75° = \cos 11.25° = \sqrt{\left(\frac{\sqrt{2+\sqrt{2}}}{2} + 1\right)/2} = \frac{1}{2}\sqrt{2 + \sqrt{2+\sqrt{2}}}.$$

And that is more pretty than it has any right to be.

For the other angle, we have $\sin 56.25° = \cos(90° - 56.25°) = \cos 33.75°$, and since $33.75 = 67.5/2$ and $\cos 67.5° = \sin 22.5° = \sqrt{2-\sqrt{2}}/2$ (again by the Origami Trigonometry activity), we have that

$$\cos 33.75° = \sqrt{\frac{\cos 67.5° + 1}{2}} = \sqrt{\left(\frac{\sqrt{2-\sqrt{2}}}{2}+1\right)/2} = \frac{1}{2}\sqrt{2+\sqrt{2-\sqrt{2}}}.$$

Indeed!

Now we are ready to use the law of sines on $\triangle EGH$. We have

$$\frac{GH}{\sin 56.25°} = \frac{GE}{\sin 78.75°}$$

and $GH = \sqrt{2} - 1$. Thus,

$$GE = (\sqrt{2}-1)\frac{\sqrt{2+\sqrt{2+\sqrt{2}}}}{\sqrt{2+\sqrt{2-\sqrt{2}}}} = 3 - 2\sqrt{2} + \sqrt{10 - 7\sqrt{2}}.$$

That last equality is obtained after a long bout of square-root slogging. Or it can more easily be simplified using a computer algebra system.

In any case, we finally have that our radius of the star ring is

$$FH = GH + FG = GH + GE = \sqrt{2} - 1 + 3 - 2\sqrt{2} + \sqrt{10 - 7\sqrt{2}}$$

$$= 2 - \sqrt{2} + \sqrt{10 - 7\sqrt{2}} \approx 0.902812.$$

Therefore, if we start with square pieces of paper with side length 1, the resulting ring made with 16 units will have radius ≈ 0.902812, measured from the center of the ring's hole to the tip of one of the points.

Obviously, this radius question is much more involved than the original problem of determining the number of units needed.

Final notes

This activity as described on the handout could be used in math classes as low as middle school, as long as students are familiar with measuring degrees of angles and know that the sum of the interior angles in a polygon with n sides is $(n-2)180°$.

To handle the radius question, students really need a solid high-school level knowledge of trigonometry and of how to muck around with square roots. There are likely many methods one could use to solve this problem; the above solution is only one way to do it. Again, feel free to allow your students to get an approximate answer to the radius problem instead of insisting for an exact solution. (Unless you really want to challenge them!)

The main questions of this activity were asked and solved by Ann Farnham, a teacher and former student in Western New England University's MAMT (Master of Arts in Mathematics for Teachers) program. The students she was working with when developing this activity were middle-school students, and for the radius question she encouraged them to just get a decimal approximation of the answer.

Activity 13
FOLDING A BUTTERFLY BOMB

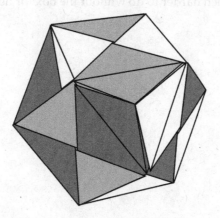

For courses: geometry, math for liberal arts

Summary

Students are taught how to make Ken Kawamura's "Butterfly Bomb" and/or the "capped octahedron" bomb model. After making it, they learn how to make it explode. In order to repeat this trick, they need to become proficient in assembling it.

Content

The Butterfly Bomb model's final shape is that of a cuboctahedron whose triangle sides have become concave, pyramid-shaped chambers. Thus the construction of this model requires becoming familiar with this object. The model is also quite hard to put together, and so students have to work at understanding the object's structure and symmetry to help them get it together. The explosive nature of the model provides motivation.

The capped octahedron version is actually a "dual" of the Butterfly Bomb, although it requires fewer pieces of paper and is easier to put together.

Handouts

All the handouts are origami instructions.

(1) Instructions for making the Butterfly Bomb.

(2) Instructions for the classic Masu Box model, which can aid in the Butterfly Bomb construction.

(3) Instructions for the capped octahedron bomb model.

Time commitment

The capped octahedron model will take 30–40 minutes. The Masu Box and Butterfly Bomb will take a full hour. The Butterfly Bomb by itself would also take about an hour, since it's so much harder to do without the box for help.

Making a Butterfly Bomb <small>(invented by Kenneth Kawamura)</small>

You'll need 12 pieces of stiff, square paper. Use 3 colors (4 sheets per color).

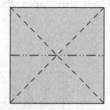

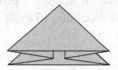

(1) Take a sheet and fold both diagonals (with valley folds). Fold in half horizontally with a **mountain** fold.

(2) Collapse all these creases at the same time to get the above figure. Press flat and score the creases firmly. Then open it up again.

Repeat with the other 11 squares.

Putting it together: The object is to make a **cuboctahedron**, which has 6 square faces and 8 triangle faces.

First form a square base using four units as shown. The units should be layered over-under-over-under to weave together.

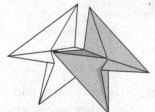  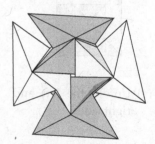

Then use a unit to make a triangle-shaped cavity to the side of the square base. Again, the units should weave. It will be **hard** to make them stay together. Working in pairs (with more hands) will help. Do this on each side of the square base.

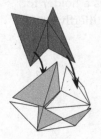 

Keep adding units, making square faces and triangle cavities. It won't stay together until the last one is in place.

Why is it a bomb? Toss the finished model in the air and smack it underneath with an open palm to see!

The Classic Masu Box

This box is a classic Japanese model. It also can be a big help for making the Butterfly Bomb. If making a Butterfly Bomb from 3 in to 3.5 in paper, then make your Masu Box out of a 10 in square.

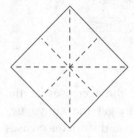 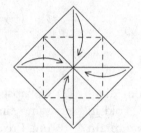

(1) Crease both diagonals and both horizontals.

(2) Fold all four corners to the center.

(3) Fold each side to the center, crease, and unfold.

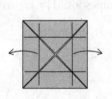 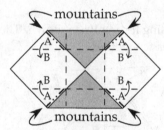 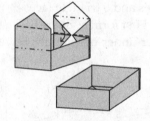

(4) Unfold the left and right sides.

(5) Use the mountain creases shown to form a 3D box. The A regions should land on top of the B regions as shown...

(6) ...here. Then fold the other sides inside, making them line up with the other tabs, to finish the box!

How this can help with the Butterfly Bomb: Use the Masu Box as a holder for the Butterfly Bomb units as you make it. The square sides of the Butterfly Bomb should be flat against the Masu Box sides.

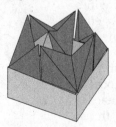 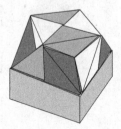

Making a Butterfly Bomb Dual

You'll need 6 pieces of square paper. Use 3 colors (2 sheets per color).

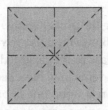

(1) Take a sheet and fold both diagonals (with valley folds). Fold in half both ways with **mountain** folds.

(2) Collapse all these creases at the same time to get the above figure. Press flat and score the creases firmly. Then open it up again.

Repeat with the other 5 squares.

Putting it together: The object is to arrange the units like the 6 faces of a cube. They should weave together to form eight pyramids

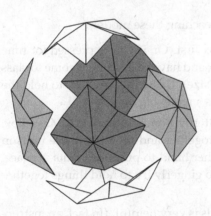

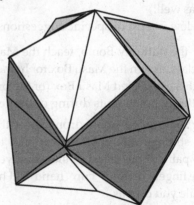

The units will not want to stay together until the last one is in place. If you have trouble, work with someone else to help. (The more hands the better!)

This model is also a "bomb." Toss it in the air and smack it from underneath with an open palm to make it explode!

Question: What does this shape remind you of? How would you describe it?

"SOLUTION" AND PEDAGOGY

Nearly the whole of this activity is in the construction of the models. While there are concepts that can be gleaned and elaborated from these models (as will be described below), it is the construction process itself that helps develop mental images and understanding of certain polyhedral shapes.

The greatest challenge with this activity is in putting the units together to make the model. Both the capped octahedron and Butterfly Bomb models are very hard to construct because they are very unstable until the last unit is inserted. Unlike the PHiZZ unit or other modular units you or your students may have seen, these units have no locking mechanism. The units basically rest upon one another, and only when they are all together will their combined weaving provide any kind of lock. In fact, these models are so delicate that usually the act of inserting the last unit will cause everything to come apart a little, requiring the whole object to be squeezed slightly to get everything in its proper place.

Instructors *must* practice these models many times before challenging a class with them. Often students will need one-on-one help to begin putting them together, and if the instructor has a hard time with these models then it probably won't go well. (Then again, trying to make these models with a math club where it's a discovery process for both the students *and* the faculty can be a great experience as well!)

Below is a list of specific suggestions for teaching these models.

- For the Butterfly Bomb, teach the Masu Box first. Or, if you're pressed for time in class, assign the Masu Box for homework and have each student come to class with a completed Masu Box (of the proper size). Using this as a tray to hold the Butterfly Bomb units during construction is a really big help.

- For the capped octahedron, or for the Butterfly Bomb without the Masu Box, the strategy should be to get 3 or 4 units together and then cup these units in the palm of one hand while using your other hand to put more units in place. The fingers of your "cup" hand will have to gingerly try to hold things together while you do this.

- With these bomb models, two pairs of hands is very helpful. (In fact, an instructor making one for the first time might want to enlist the help of a colleague.) Some students will finish these models much faster than others, and these fast students should help their neighbors with their models. This makes the instructor's job easier and helps foster student collaboration.

- The pictures on the handouts were designed to be both efficient and pedagogically meaningful. Not only would it take much more paper to show, step-by-step, how to assemble the capped octahedron, but it would harm the educational experience as well. Students need to mentally visualize what is going on and *then* experience it by putting the units together. But this also means that the instructor will have to do a lot of one-on-one assisting of students until it "clicks" in their heads.

The time required to teach these models will vary depending on a number of factors. The capped octahedron model can take 30–40 minutes for everyone to make one, explode it, and reassemble it. The Butterfly Bomb will take longer, needing at least a full hour. If the Masu Box is taught first, this will make the assembly much easier and quicker, but the total time devoted to the activity will be the same (about 15–20 minutes for the Masu Box and 40 minutes for the Butterfly Bomb).

Emphasizing content

Simply making these models will latently teach the students much about the structure of certain polyhedral shapes, but emphasizing the connections afterward will do much to reinforce it.

Capped octahedron. As the handout suggests, this model can be viewed as each piece of paper being a side of a cube. In fact, if you took a cube and "dented" the edges by pressing in on their midpoints you can get this very same object.

However, the finished object looks more like a bunch of pyramids. In fact, when making the object it's often useful to build one of these pyramids at a time so as to keep track of one's progress as units are added. So, instructors should ask their students to count how many of these pyramids are in the final model—there are eight. And what geometric figure is made by the base of these pyramids? An equilateral triangle. What famous object is made up of eight equilateral triangle faces? The octahedron! Thus we can think of this model as having an octahedron inside it, where the eight pyramids are capping each face of the octahedron. This is why I refer to this model as a "capped octahedron" and why I do not use this moniker in the handout. I prefer to let students build the model and then discover what properties it has. But if students are already familiar with the octahedron, telling them at the outset that the shape they'll make is a capped octahedron may help them put the thing together.

Also, if the concept of duality has been introduced in your class, then it should make perfect sense for students to visualize this model from the dual perspectives of the cube and the octahedron.

Note: This capped octahedron shape can be made from many different origami units. In fact, the Sonobè Unit [Kas87] is a very popular unit, 12 of which can make this shape (but with a different coloring pattern). Some of your students may have previously made this model. Many origami books and references refer to this shape as a *stellated* octahedron, but this is incorrect. Stellation means to extend each face of the polyhedron until the face planes intersect in interesting ways. Doing this to the octahedron does result in "caps" being placed on the triangular faces, but they will be perfectly regular tetrahedra on each face, not the right triangle pyramids that we see in our model. Thus, students should not be encouraged to refer to this model as a stellation.

Butterfly Bomb. The basic structure of this model is a cuboctahedron whose triangle faces have been dimpled to become pyramid-like chambers. (It is close to, but not exactly, a *cubohemioctahedron* [Wei1], which is a cuboctahedron whose triangle faces have become regular tetrahedral cavities. In the Butterfly Bomb they are right-angle tetrahedra cavities.) Students who make this model inside a Masu Box might prefer to view it as a cube-like shape whose corners have been truncated at the midpoints of each edge. (The square faces of the model represent the faces of the former cube, and the vertices of the cuboctahedron are the midpoints of the edges of the former cube.) But this model can also be viewed as an octahedron whose corners have been truncated at the midpoints of the edges. This is another demonstration of the duality between the cube and the octahedron.

Both of these models also have a left- or right-handedness, depending on how the units are woven together. (For example, does the weaving on each square face of your Butterfly Bomb go clockwise or counterclockwise?)

Packing to fill space. One very surprising fact is that these Butterfly Bomb and capped octahedron models can tessellate three-dimensional space (see the figure below). This is really just a consequence of the fact that three-dimensional space can be tessellated with octahedra and cuboctahedra, and the caps from the capped octahedra fit perfectly into the pyramid cavities of the Butterfly Bomb.

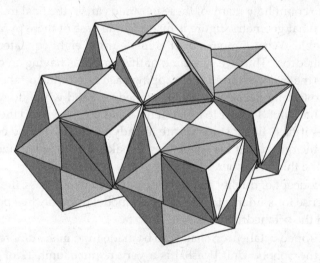

Students usually find this tiling property very exciting. Although using class time to do both models may be excessive, instructors could do one in class and assign the other for homework. Then students can be encouraged to discover this three-dimensional tiling property on their own.

Variations

There are a number of variations that can be made from these two models. For example, suppose that we reverted the pyramid cavity "dimples" in the Butterfly Bomb so that they poked out of instead of into the model. Then, the basic shape of the model would be a cube, as shown below on the left. The "units" for this model are merely squares folded in half. This model is *very* unstable.

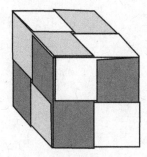 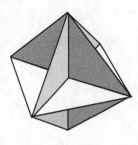

Alternatively, we could invert the pyramids on the capped octahedron to be poking inside the model. This turns out to be equivalent to flipping the units "inside out," and the result is an *octahedral skeleton*, shown in the right figure above. This model is very stable and is not a "bomb" at all. In fact, we will see this octahedral skeleton structure again in the next activity, Molly's Hexahedron (which itself is similar in structure to the models in this activity).

These and other variations were discovered by a variety of origamists including Robert Neale, Lewis Simon, Kenneth Kawamura, and Michael Naughton. (Although Kawamura is credited as being the first to capitalize on the instability of these models as a way to make them explode.) In fact, there is an entire continuum of models (several continua, in fact) between the six-piece octahedron skeleton shown above and a six-piece cube (not shown) made from squares where the four corners have been folded to the center. Students and instructors should feel free to explore such variations themselves.

Activity 14
Molly's Hexahedron

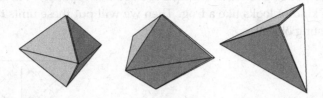

For courses: precalculus, geometry, math for liberal arts

Summary

A very simple, 3-piece modular object is taught that takes the form of an unusual-looking hexahedron. Students are then asked, "If the starting square pieces of paper that we used have side length one, what is the volume of the finished object?" Once this volume is computed, students are asked to reconcile their answer with how big a $1 \times 1 \times 1$ cube would be.

Content

At the most basic level, this activity is about computing and understanding volumes. The only volume formula needed for this is the volume of a pyramid formula, which is introduced to students as early as middle school. Interpreting the answer as how many of these hexahedra should fit inside a $1 \times 1 \times 1$ cube is, however, more challenging and makes for a great activity in college-level geometry classes, since it leads to the concept of polyhedral dissections. Therefore this activity is scalable to many different levels of math classes.

Handouts

There are two handouts. The first one describes how to make the hexahedron and asks students to find the volume. The second is optional, showing how to make the origami octahedron skeleton, which fits together with Molly's Hexahedron very nicely.

Time commitment

Teaching the unit and assembly of the hexahedron is pretty fast. Most students will finish it in under 10 minutes. Computing the volume can be a tricky puzzle, but is fundamentally simple; 30 minutes for the whole activity is easily possible, perhaps less or more depending on how in-depth instructors want to explore the $1 \times 1 \times 1$ cube dissection aspect.

Molly's Hexahedron

This model, invented by Molly Kahn, requires 3 squares of paper. Fold each square into a unit that kind of looks like a frog. Then we will put these units together to make an interesting object!

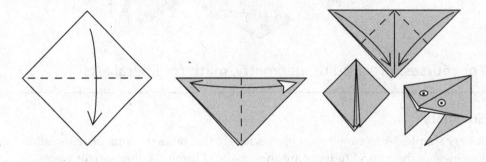

(1) Fold a diagonal.

(2) Fold in half. Unfold.

(3) Fold the corners to the bottom and you are done! Make 2 more.

Putting them together:

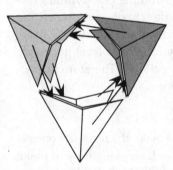

Slide the "legs" of one frog unit into the "mouth" of another one. To make this work, the frogs need to be positioned properly, like in the left drawing. Add a third frog to complete a triangle, and squeeze them all together!

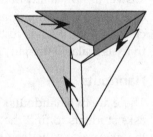

Question 1: How would you describe this object? What is the shape of its faces and how many are there?

Question 2: Suppose the side length of your original squares is 1. Then what is the volume of the finished object? Hint: Use the fact that the volume of a pyramid is $V = \frac{1}{3}Bh$, where B is the area of the base and h is the height.

The Octahedron Skeleton

This is a classic modular origami model. It was invented by Bob Neale in the 1960s and requires 6 sheets of square paper.

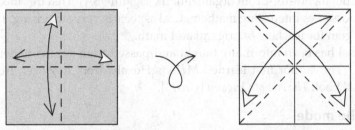

For each piece of paper, valley fold in half from left-to-right and from top-to-bottom. Then **turn the paper over** and valley fold both diagonals. It is important to turn the paper over in between doing the "horizontal-vertical" folds and the diagonal folds.

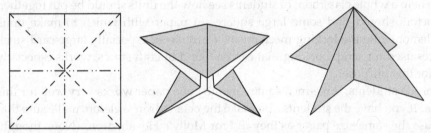

Then collapse the paper into a star shape, as shown above. The shape that results, which is called the *waterbomb base* by origamists, should have the four original corners of the square becoming long, triangular flaps.

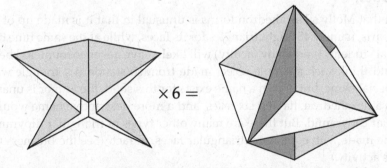

Make 6 of these waterbomb bases, 2 each of 3 colors. Then the **puzzle** is to lock them together to make an octahedron skeleton!

Hint: The triangle flaps will insert into the triangle flaps of other units in an over-under-over-under pattern.

SOLUTION AND PEDAGOGY

Molly's Hexahedron is, perhaps, one of the simplest three-dimensional modular origami models in existence. It was invented by the late Molly Kahn, whose mother was the famous origami enthusiast and catalyst Lillian Oppenheimer (one of the pivotal people in making origami as popular as it is today, and who helped found the national, not-for-profit organization OrigamiUSA). That this model is so simple and possesses interesting mathematical aspects is very surprising. It makes the model a gem for teachers of origami and math.

This model has been informally taught and passed from person to person for many years. The author first learned Molly's Hexahedron, however, from Gay Merrill Gross' book *The Art of Origami* [Gro93].

Teaching the model

The units for this model are very simple, and it will take students less than a minute to make each unit. Assembling them is a bit tricky and will take students some fiddling to figure out. But with only three units, everyone will eventually be successful, especially if students help each other out in groups.

To help a whole classroom of students see how the units should be put together, instructors should find some large squares of paper with which to make units and demonstrate the locking mechanism. Card stock, especially large card stock squares used for scrap booking that can be found in craft stores, works especially well for this purpose.

For the students, three-inch square memo cube paper works very well for this model. If you have the students also make the octahedron skeleton, make sure that they use the same size paper as they did for Molly's Hexahedron. (Note, though, that folding the octahedron skeleton takes longer, probably 20 minutes of class time.)

The solution

The solid that Molly's Hexahedron forms is unusual in that it is made up of very familiar parts, namely 45° right triangles for its faces, while at the same time being a solid that students (and many faculty) will likely have never encountered before. One would think such a simple object made from so standard a triangle would have a special name, but if such a name exists for this solid, the author is unaware of it. It is a hexahedron, having six faces, and a more descriptive name would be a triangular dipyramid. But there are many other types of triangular dipyramids; one can be made with equilateral triangular faces, in fact. (See the Business Card Modulars activity.)

Therefore, the purpose of Question 1 is merely to make the students examine this object more closely and notice that it has 6 sides, each of which is a 45° right triangle.

The volume problem in Question 2 can be attacked in a variety of ways, and depending on the class students could be encouraged to try finding the volume in a different way than the handout suggests. However, using the volume formula for a pyramid is the simplest way.

To use this formula, students need to visualize how to split the hexahedron into two pyramids, where this split happens along the equilateral triangle "equator" of the object. Students also need to figure out how the various side lengths of the hexahedron correspond to line segments in the original square, and this may require unfolding a spare unit and comparing with the finished model. The figure below illustrates these points.

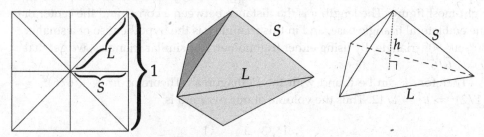

Therefore, to find the volume of the hexahedron, one needs to find the volume of one of the pyramid halves and multiply the result by two. Almost all students immediately consider trying to apply the pyramid volume formula in the configuration shown in the above right-most figure, where the base of the pyramid is the equilateral triangle face and the height extends from the center of this equilateral triangle up to vertex surrounded by the three right angles. This is **not** the easiest approach! But it is the most common, and instructors should be aware of the difficulties here.

In fact, applying the pyramid volume formula in this way easily puts the problem difficulty level into the realm of a high-school or college geometry class (or perhaps a precalculus class that utilizes geometric modeling).

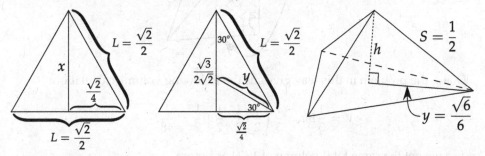

The above figures show the required reasoning needed to find the area of the equilateral triangle base of the pyramid and the height h. Note that from the above crease pattern figure, the side lengths of the hexahedron are $S = 1/2$ (since it is half the side of the square) and $L = \sqrt{2}/2$ (since it is half a diagonal). Thus the side of our equilateral triangle base is $L = \sqrt{2}/2$, and if we let x be the height of this triangle, we get that $x = \sqrt{3}/(2\sqrt{2})$, which can be found either by applying the

Pythagorean Theorem or by using the fact that x is a leg of a 30°-60°-90° triangle (with hypothenuse L and other leg $L/2$).

Therefore, the area of the base of our pyramid is

$$B = \frac{1}{2}\frac{\sqrt{2}}{2}\frac{\sqrt{3}}{2\sqrt{2}} = \frac{\sqrt{3}}{8}.$$

The height h is more difficult to figure out. It is a leg of a right triangle whose hypothenuse is $S = 1/2$ and whose other leg we label y, as shown in the previous right-most figure. The length y is the distance between a corner and the center of the equilateral triangle base, and in this triangle y is the hypothenuse of a smaller 30°-60°-90° triangle. By using either trigonometry or similar triangles, we get that $y = \sqrt{6}/6$.

Therefore h can be found with the Pythagorean Theorem: $h^2 + (\sqrt{6}/6)^2 = (1/2)^2 \Rightarrow h^2 = 1/12$. Thus the volume of our pyramid is

$$V = \frac{1}{3}\frac{\sqrt{3}}{8}\frac{1}{\sqrt{12}} = \frac{1}{48}.$$

And so the volume of Molly's Hexahedron is 1/24!

Of course, a *much* easier way to find this volume is to use as the base of the pyramid not the equilateral triangle but instead one of the 45° right triangles. As shown in the below figure (which is the same pyramid, just rotated), doing this gives us a height of 1/2 and a very easy-to-compute base volume.

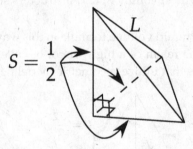

Doing the problem in this way gives us the following volume calculation:

$$V = \frac{1}{3}Bh = \frac{1}{3}\left(\frac{1}{2}\cdot\frac{1}{2}\cdot\frac{1}{2}\right)\frac{1}{2} = \frac{1}{48}.$$

And so we get the same total volume of 1/24 as before.

As one can see, the difficulty level of this problem is very scalable. The second method shown here is very much at the level of middle-school students who are learning the pyramid volume formula. Although, such students should be encouraged to carefully choose which side of the pyramid they want to be the base! For such an audience this model makes a great hands-on application of the volume formula.

For high-school or lower-level college audiences of students, the first approach described here is actually fantastic practice for using similar triangles, the Pythagorean Theorem, and trigonometry in a hangs-on problem. In the author's experience, this hexahedron object is so unfamiliar to students that only those with mathematical or geometric creative streaks will think to use a side other than the equilateral triangle side as the base of the pyramid. But instructors can also simply stipulate that the students use the equilateral triangle side as the base in order to have them practice applying trigonometry, similar triangles, and such in this problem, leaving the easier solution as a happy happenstance if a creative student thinks of it.

Interpreting the volume and dissections

Once the volume of 1/24 has been determined for Molly's Hexahedron, it is a very natural question to ask, "Does this volume make sense? Doesn't it seem low?" Indeed, if students have any extra unfolded pieces of their square paper handy, it is very easy to put some of these pieces of paper edge-to-edge to get a sense of how big a $1 \times 1 \times 1$ cube would be. If we compare the size of such a cube with one of Molly's Hexahedra, it does seem puzzling that 24 of these hexahedra should be able to fit inside this cube.

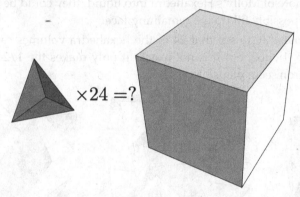

Still, the math is correct; 24 of the Molly's Hexahedron volumes will fit inside a $1 \times 1 \times 1$ cube. Seeing how this might be plausible is an interesting challenge.

One way is to use the fact that Molly's Hexahedron has all these right angles on its faces. In fact, two of its vertices are exactly like a corner of a cube, and so it makes sense to imagine putting one of Molly's Hexahedron in each of the 8 corners of our $1 \times 1 \times 1$ cube. And since the side length $S = 1/2$, these 8 hexahedra will fit perfectly in such an arrangement, as shown in the below-left picture.

That puts 8 of the hexahedra into our cube. More can be nicely placed inside if we again split our hexahedra into two right angle pyramids. If we do this to two of Molly's Hexahedra and rearrange them as shown in the below-center picture, we can make a square-based pyramid whose base fits perfectly into the remaining space in the center of each face of our cube. The below-right picture shows two of these square-based pyramids placed in the top and bottom faces of our cube, and one can see that these pyramids meet in the cube's center.

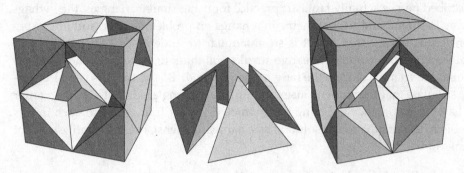

Therefore, filling in the center of each face of the cube in this way adds two hexahedra per face into our $1 \times 1 \times 1$ cube. That adds $2 \times 6 = 12$ hexahedra to our 8 at the corners, and we are up to 20 hexahedra that have been placed inside out cube.

Now, looking carefully at the above-right figure, one can see that the center-placed square-based pyramids do not touch face-to-face the hexahedra placed at the corners. So there is still room left inside our cube, and it is conceivable that if we turned four more of Molly's Hexahedra into liquid, they could be poured into these spaces and possibly fill up the remaining face.

This is only one way to see that 24 of the hexahedra volumes can be placed inside the $1 \times 1 \times 1$ cube, but it is not exact. It only makes the 1/24 volume of Molly's Hexahedron seem plausible.

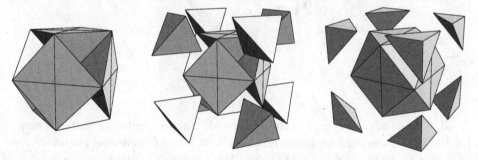

The above pictures, however, offer an exact dissection of the cube corresponding to attempted hexahedron packing. If we were to take the square-based pyramids (each made from four of our right triangle pyramids, each of which is half of a hexahedron) placed at the center of each square face of the cube and look at these with no other hexahedra added, we would get the figure shown above-left. This looks like a cuboctahedron with dimples at all the triangle faces, and this may remind readers of the Butterfly Bomb activity in this book. But this is *not* the same object as the Butterfly Bomb. The sides of the dimpled cavities of our object are equilateral triangles, whereas in the Butterfly Bomb model they are 45° right triangles.

If we fill these dimpled cavities with regular tetrahedra, as illustrated in the above-center picture, then we will have a perfect cuboctahedron solid. Then if we place copies of our right triangle pyramids (half of Molly's Hexahedron) on each of the triangle faces of this cuboctahedron, we will have completely filled our $1 \times 1 \times 1$ cube. This last step is shown in the above-right picture. (Please note that while it would be nice to have a way to fold the regular tetrahedra in this dissection with a unit as simple as that for Molly's Hexahedron, this is not easily done. Folding a regular tetrahedra requires a move along the lines of those found in Activity 1, let alone trying to get it to be the proper size to fit in this dissection.)

Let us count the pieces and their volumes of the cube dissection just described. We have 8 half-hexahedra at the corners, 8 regular tetrahedra underneath those, and the same central structure as before, made of $4 \times 6 = 24$ half-hexahedra (or $2 \times 6 = 12$ hexahedra, if you prefer). That is 32 half-hexahedra, or 16 full copies of Molly's Hexahedron, giving us a volume of $16 \times 1/24 = 2/3$ so far.

The regular tetrahedra in this dissection have side lengths equal to $L = \sqrt{2}/2$. The height h of this tetrahedron is a leg of a right triangle with hypothenuse L and other leg y, which is the same length $y = \sqrt{6}/6$ that we calculated during our first attempt at finding the volume of Molly's Hexahedron (shown in the bottom-right figure on p. 143). So by the Pythagorean Theorem, we have

$$h^2 = \left(\frac{\sqrt{2}}{2}\right)^2 - \left(\frac{\sqrt{6}}{6}\right)^2 \Rightarrow h = \frac{1}{\sqrt{3}}.$$

And so the volume of the regular tetrahedra in this dissection is

$$\frac{1}{3}\left(\frac{1}{2} \cdot \frac{\sqrt{2}}{2} \cdot \frac{\sqrt{3}}{2\sqrt{2}}\right)\frac{1}{\sqrt{3}} = \frac{1}{24}.$$

Ah ha! These regular tetrahedra have the same volume as Molly's Hexahedron! Since we have 8 of these in our dissection, they add $8 \times 1/24 = 1/3$ to the volume. When added to the hexahedra in the dissection, we have that the volumes $2/3$ plus $1/3$ give us the total volume 1 of the cube.

There is a geometric explanation for why the volume of Molly's Hexahedron is the same as that of the regular tetrahedra in our cube dissection. The reason has to do with the natural way in which a tetrahedron can be inscribed in a cube. If one takes four vertices of a cube with the property that no two of them are connected by an edge, then these four vertices form the corners of a regular tetrahedron contained inside the cube. If one takes the volume contained in this tetrahedron and removes it from the cube, the remaining parts of the cube will be made up of four copies of Molly's Hexahedron halves (the right triangle pyramid we obtained by splitting Molly's Hexahedron in half), each touching the others along an edge.

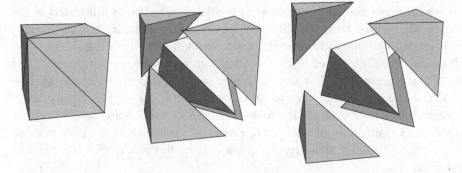

The above set of pictures illustrates this. Another way to think about it is that if we were to take one of the regular tetrahedra in our cube dissection and place Molly's Hexahedron halves on each of its four equilateral triangle faces, we would get a perfect cube.

Now, readers are invited to verify on their own that if one were to take a cube and compute the volume of this inscribed tetrahedron, the result would be that the tetrahedron's volume is exactly 1/3 that of the cube. (Take a $1 \times 1 \times 1$ cube, and the inscribed tetrahedron will have side length $\sqrt{2}$. Then use the same tetrahedron calculation we did previously.) The four Molly's Hexahedron halves make up the remaining 2/3 volume of the cube. Two of these Molly's Hexahedron halves could, of course, be put together to make one Molly's Hexahedron, and thus the volume of these two halves put together would be 1/3 the volume of the cube. That is, the volume of the tetrahedron is the same as that of Molly's Hexahedron.

More on pedagogy

Instructors who have read this far may be thinking, "How on Earth could I expect my students to come up with that complicated dissection?" However, this question misses the point, in a sense. The point of the above exposition was to provide evidence for *how rich the mathematics of this simple model is*. The relatively simple task of computing the volume of Molly's Hexahedron and questioning how this volume makes sense when compared to the volume of a $1 \times 1 \times 1$ cube can lead to a very in-depth exploration of a non-trivial dissection of the cube and the cuboctahedron, which further leads to the classic way to inscribe a regular tetrahedron in a cube. Such explorations in how polyhedra can inscribe or fit into each other are fascinating but often hard for students to visualize, let alone discover for themselves.

Explorations of how, say, a tetrahedron can fit inside a cube, or the relationships between the cube, the cuboctahedron, and the octahedron, can all fit into curricula on the Platonic solids and three-dimensional solids in general, or any unit on 3D solid geometry. Cromwell's book *Polyhedra* [Cro99] is an excellent resource on polyhedral geometry, its history, and the relationships between the various 3D solids.

Teachers who have the time and desire to have their students fully explore these dissection aspects of Molly's Hexahedron might find it useful to teach their students how to make *half of* a Molly's Hexahedron out of origami. One easy way to do this is to simply invert one half of the hexahedron into the other half, but this leaves a cavity for the equilateral triangle side. Still, it does produce the basic right triangle tetrahedron shape used in the above dissection illustrations.

Another way to make this shape is by modifying the unit for Molly's Hexahedron as follows:

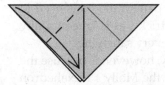 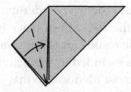 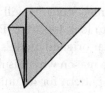

(1) Take a Molly's Hex-
ahedron unit and re-
fold the left arm.

(2) Fold the lower-left
edge to the center
line. *Make this crease
sharp!*

(3) That's it!

To join these units, put the other half of each unit together as was done for Molly's Hexahedron. This will leave the extra folded flaps sticking out, as shown below. These flaps should be layered, over-under-over, to make the equilateral triangle side. Note that there will be a rather attractive triangle hole in the middle of that face.

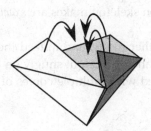 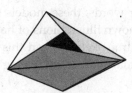

Multiple copies of this half-hexahedron model can be put together to mimic the cube dissection described previously. However, in order to make such a dissection stay together in a cube form, either a box is needed to hold the pieces in place, or multiple people are need to hold everything together, basically to form a supporting box with their hands.

Playing with these half-hexahedra is a sure-fire way to allow students to see fully how 24 copies of Molly's Hexahedron can fit inside a cube. But it does require a lot of folding.

Bonus problem: octahedron volume

The Octahedron Skeleton model is a fun one to teach and fold in itself. It's a classic in the modular origami cannon, and many other similar models have spun off it. It was invented by the legendary paper folder Robert (Bob) Neale, who is also known for his contributions to the professional magic community, as well as for having been a professor of theological psychology for many years.

Note that the assembly of this octahedron model is left as a puzzle on the handout. In the author's experience, all students need to see is an example of this model. When allowed to examine such a model up-close and then play with their own units, students will figure out the over-under-over pattern the units need in order to lock together. Putting the last unit in place is tricky, as is true in many modular origami models. But the results for this one are very satisfying.

The reason why this model is included in this activity, however, is because the crease pattern for its units is almost identical to that of the Molly's Hexahedron units. These crease patterns are shown below, with the octahedron skeleton unit to the left and that for Molly's Hexahedron to the right.

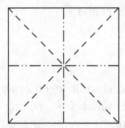 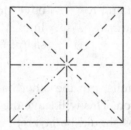

Notice that the only difference between these crease patterns is their mountains and valleys. Everything else is the same. Why is this significant? Because it means that the kinds of triangles that Neale's octahedron skeleton makes are *exactly* the same triangles as those in Molly's Hexahedron.

In other words, these models should fit together nicely, and indeed they do! Below is shown illustrations of how Molly's Hexahedron can fit snugly on top of the octahedron skeleton, and this can be extended with multiple copies of these models to make rather interesting "towers."

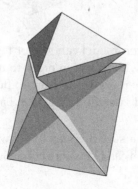 

When a classroom of students make these hexahedron and octahedron skeleton models, one cannot miss the opportunity to allow students to put them together in such towers.

Of course, this then begs the question, "What is the volume of the octahedron shape that this octahedron skeleton makes?" If the volume of Molly's Hexahedron has previously been computed, then finding the volume of the octahedron is easy. Eight half-hexahedra fit inside this octahedron, and so its volume must be $8 \times 1/48 = 1/6$. Again, this seems very small when compared to the $1 \times 1 \times 1$ cube! But this is just another example of how volumes can be so difficult to estimate in comparison to one another.

Bonus bonus problem: different crease patterns. Seeing this octahedron skeleton model leads to yet another question. Since the same crease pattern with two different mountain-valley assignments led to two very different models, are there any other ways to pick the mountains and valleys in this crease pattern to make other objects? And might these other objects also fit together with the octahedron and Molly's Hexahedron well?

The answer to this question is, "Yes!" However, we will leave it to the reader to fully explore. Be aware, though, that there is another model in this book that offers one answer to this problem.

Activity 15
BUSINESS CARD MODULARS

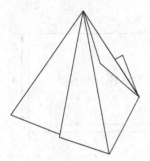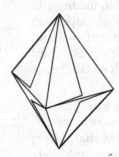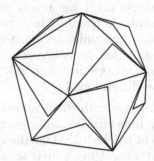

For courses: geometry, math for liberal arts

Summary

Students are shown a very simple modular origami unit that is made from business cards and asked to explore the kinds of objects that can be made with it.

Content

This unit can make any polyhedron with all triangle faces and no vertices of degree 6 or higher. Thus, at a basic level this activity is about exploring such polyhedra, starting with the regular cases of the tetrahedron, octahedron, and icosahedron and moving into other solids like the triangular dipyramid and snub disphenoid.

Handouts

(1) Describes how to make the basic unit and challenges the students to make various polyhedra with it.

(2) An optional handout with pictures of Johnson solids with all triangle faces.

Time commitment

Teaching the unit and having students make the tetrahedron and octahedron will take a good 30–40 minutes. The icosahedron or other objects will take longer, of course. The whole project could be spread over a few class days, or some of the models could be left for homework or out-of-class excursions.

Business Card Polyhedra

Business cards are a very popular medium in **modular origami**, where pieces of paper are folded into **units** and then combined, without tape or glue, to make various shapes. Standard business cards are 2 inch × 3.5 inch rectangles, or have dimensions 4 × 7.

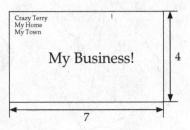

Below are instructions for making a very simple unit from business cards that can make many different polyhedra. **Make the creases sharp!** This unit was originally invented by Jeannine Mosely and Kenneth Kawamura.

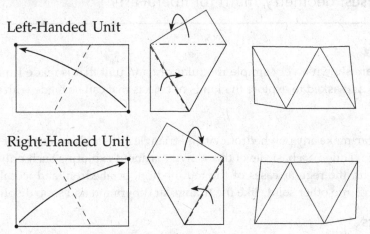

Left-Handed Unit

Right-Handed Unit

Question 1: Notice that these simple folds on a business card give us, it seems, equilateral triangles. Are they **really** equilateral? How can we tell?

Task 1: Make one left- and one right-handed unit and find a way to lock them together to make a **tetrahedron** (shown below left). After you do that, use 4 units to make an **octahedron** (shown below right). We're not telling you how many left and right units you need—you figure it out!

Task 2: Now make 10 units (5 left and 5 right) and make an **icosahedron** with them. An icosahedron has 20 triangle faces. (See the below figure.) Putting this together is quite hard—an extra pair of hands (or temporary tape) might help.

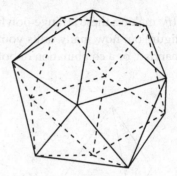

Task 3: What other polyhedra can you make with this unit? Hint: There are lots more. Try making something using only 6 units. How about 8 units? Try to describe the polyhedra that you discover in words.

Johnson Solids with Triangle Faces

Try making these strange polyhedra using the business card unit. You'll have to figure out how many units you'll need and whether they should be left- or right-handed, or a combination of both!

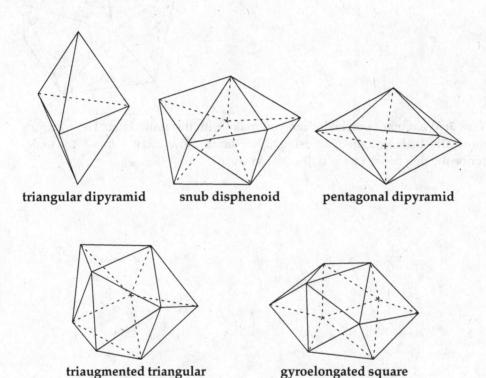

triangular dipyramid   **snub disphenoid**   **pentagonal dipyramid**

triaugmented triangular prism

gyroelongated square dipyramid

SOLUTION AND PEDAGOGY

Question 1

It seems entirely a coincidence that business cards produce equilateral triangles so well. But the 4×7 dimensions work because $\arctan(4/7) = 29.7\ldots° \approx 30°$.

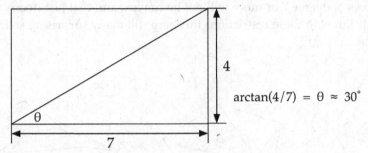

$$\arctan(4/7) = \theta \approx 30°$$

The Tasks

These units lock together entirely by a "hugging" mechanism. The short flaps wrap around and "hug" the sides of other units, holding them in place. This only works if the creases are *made sharp*, so you should emphasize this. Sharp creases can be effectively made by running over each fold with a ruler or flat pen. The tetrahedron goes together the easiest, where the left- and right-handed units grasp each other like a pair of hands. There are several ways to make an octahedron from four units; one can use either 2L and 2R units, or all L, or all R.

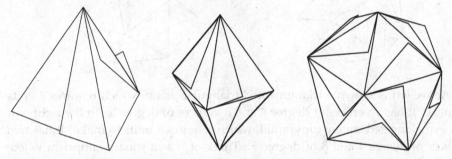

The icosahedron is very hard to put together, but only because the units want to fall apart until the very last one is inserted. Having pairs of people work together to make this is a very good idea, and you might want to have adhesive tape available for extra help. Once the model is together it's fairly sturdy, but one shouldn't squeeze it too tightly!

While students may already be familiar with the Platonic solids with all triangle faces (and if they're not, this activity will help fix that!), they will most likely have to think hard to come up with irregular polyhedra with all equilateral triangle faces. However, there are *many* such irregular polyhedra, and most require no more than six to ten business card units to make. The fact is that this business card unit can make *any* polyhedron having

(1) all equilateral triangle faces and

(2) all vertices of degree 5 or less.

The reason for (2) is that if you have a vertex of degree 6, then the equilateral triangle faces around it make a flat plane, and the units won't be able to hug each other. Vertices of degree 7 or more will not be convex, and that just doesn't work for this unit. But with these restrictions there are still many surprising solids that can be made.

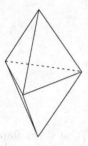 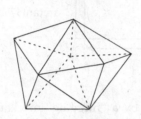 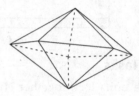

For example, above are shown pictures of a triangular dipyramid (left, 3 units), a snub disphenoid (center, 6 units), and a pentagonal dipyramid (right, 5 units).

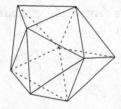 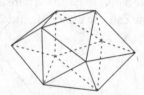

Above left is shown a triaugmented triangular prism, which requires 7 units to make. It has 3 vertices of degree 4 and 6 vertices of degree 5. To the right of it is a gyroelongated square dipyramid, which requires 8 units to make. It has two vertices of degree 4 and 8 of degree 5. Think of it as a square antiprism where square pyramids have been placed on the square faces. You could try to make a gyroelongated triangular dipyramid, but this doesn't really exist as a solid because some of the triangle faces become flat planes, resulting in a parallelepiped. (You can try to make it anyway using business cards, but it doesn't stay together very well.)

These are all examples of Johnson solids, a family of convex polyhedra having all sides regular polygons with equal edge lengths, excluding the other traditional families of polyhedra (the Platonic solids, Archimedean solids, prisms, and antiprisms). See http://www.mathworld.com or the great graphics program Poly (which can be downloaded from the web at http://www.peda.com/poly/) for more information.

Pedagogy

Fundamentally, all this activity is doing is giving students a chance to construct a variety of polyhedra. That may not seem like much, but the pedagogical value of such exercises should not be underestimated. There is a long tradition, probably going back to the Greeks and maintained today by such luminaries as Magnus Wenninger [Wen74] and George Hart [Hart01], of constructing polyhedra as a way to develop understanding of spatial relations and geometry. In fact, many people fail to really get a sense of what, say, the Platonic solids are until they actually make one with their own hands. Holding a pre-made model is not enough! The student needs to build one with her own hands to get a deeper sense of the nature of these objects.

The business card unit offers one way to do this, and the fact that it uses such an everyday object makes this surprisingly fun. Each unit covers two adjacent triangle faces of the polyhedron under construction, so students need to keep track of things like the degrees of the polyhedron's vertices as they proceed.

Depending on their manual dexterity and three-dimensional visualization skills, students will have very mixed success with this activity. Some will quickly assemble the tetrahedron and octahedron, while others will need lots of help getting the tetrahedron together. Having students working in groups can even things out, where fast students will stop to help the slower ones along. This is all the better anyway because the icosahedron is much easier for students to do in groups.

Students will be very unlikely to discover very many of the Johnson solids on their own. Someone may come up with the triangular and pentagonal dipyramids, but the others are just not very intuitive. Once students are convinced that there are no other possibilities than the ones they've come up with, then it is time to unveil some of the more complicated Johnson solids.

This is what the second handout is for, but this is entirely optional. If computer projection facilities are available in class, instructors can project pictures of these solids from the MathWorld web page or from the Poly program (as mentioned previously). Or instructors can assign their students to find more polyhedra to make with business cards on their own. With the web at their disposal, it is totally reasonable to expect students to come up with the triangle-faced Johnson solids for homework.

This activity can also be useful for reinforcing a variety of concepts in polyhedra or planar graph theory. Students in a math for liberal arts class are often exposed to Euler's formula $V - E + F = 2$, for example, which can be verified with these business card polyhedra.

On obtaining large supplies of business cards

Business cards make up a subfield of modular origami. Many other business card units can be found by web searching. (Also see the Modular Menger Sponge activity in this book.) Those who delve into this area will discover that not all business cards are the same. While they all have the same dimensions, the quality of the card stock can vary quite a bit. Some cards will have a glossy coating that may crack when folded. Others may be of a slightly less weight than standard and fold more easily.

Obtaining lots of cards can sometimes be easy. Visit an office supply, photo-copy, or printing store that prints business cards for customers and ask if they have any discarded cards. Often printing errors or flaky customers result in boxes of cards being left unwanted. Such stores are often happy to give these away. Blank cards may also be purchased, but it can be fun to collect random cards to see what the printing will reveal when folded. Business card folding enthusiasts often collect cards from restaurants and businesses, sometimes sorting them by color so that they may be used to artistic effect.

In fact, one can assign students to collect business cards of their own and bring a supply of 10–20 cards with them to class. As long as you give them advanced notice, this is entirely reasonable (although you'll still want a supply of cards on hand to help those in need).

Activity 16
FIVE INTERSECTING TETRAHEDRA

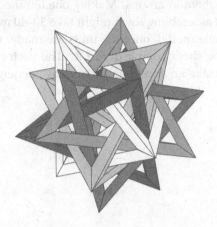

For courses: geometry, math for liberal arts, calculus, multivariable calculus

Summary

Students learn how to make a modular origami tetrahedral frame using Francis Ow's 60° unit. Then, they are challenged to, in groups, weave five such tetrahedra together to make an origami version of the compound of five tetrahedra.

Content

Making one tetrahedral frame is not hard. But weaving five together in the proper way is a big puzzle! To do so requires grappling with some unusual symmetries in three dimensions that are based on natural properties of the dodecahedron.

As a possible follow-up, finding the optimal "strut width" for this model is a challenging vector geometry and calculus problem. It requires serious study of the complicated symmetries of the model and is not something that is easy to do without a physical model in one's hands.

Handouts

(1) "Five Intersecting Tetrahedra" (2 pages) and "Linking the Tetrahedra Together" (1 page) describe how to make the model.

(2) "What Is the Optimal Strut Width?" (2 pages) leads students through the basic steps of a vector geometry and calculus solution to the problem.

161

Time commitment

While the units are simple to make, there are 30 of them, so folding all the units might take an individual over 30 minutes. This probably should not be done in class unless students do them in groups. Making one tetrahedron is not so bad— but folding the units *and* assembling them might take 30–40 minutes.

Constructing the whole model, once the units are made, would take another 30–40 minutes due to the sheer complexity of it. The vector geometry–calculus activity would probably take an additional 20 minutes, depending on the level of the students.

Five Intersecting Tetrahedra

This origami model is a real puzzle! But first we'll start with the **one tetrahedron** made from Francis Ow's 60° unit [Ow86].

Francis Ow's 60° unit

This requires 1×3 paper. So fold a square sheet into thirds and cut along the creases.

 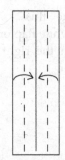 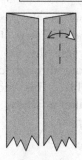

(1) Crease in half lengthwise.

(2) Fold the sides to the center.

(3) At the top end, make a short crease along the half-way line of the right side.

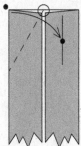 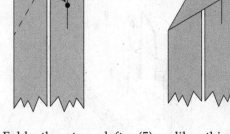 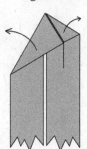

(4) Fold the top left corner to the pinch mark just made **and at the same time** make sure the crease hits the midpoint of the top…

(5) …like this. Fold the top right side to meet the flap you just folded.

(6) Undo the last two steps.

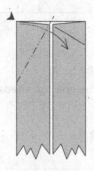 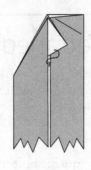 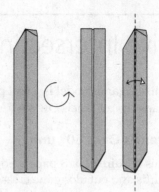

(7) Now use the creases made in step (4) to **reverse** the top left corner through to the right. This should make a white flap appear...

(8) ...like this. Tuck the white flap underneath the right side paper.

(9) Now rotate the unit 180° and repeat steps (3)–(8) on the other end. Then fold the whole unit in half lengthwise (to strengthen the spine of the unit) and you're done!

Locking the units together: Three units make one corner. **Make sure** to have the flap of one unit **hook** around the spine of the other!

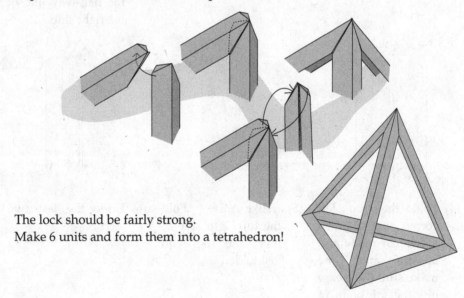

The lock should be fairly strong.
Make 6 units and form them into a tetrahedron!

Linking the Tetrahedra Together

The five tetrahedra must be woven together, one at a time. The second tetrahedron must be woven into the first one as it's constructed. That is, it's not very practical to make two completed tetrahedra and *then* try to get them to weave together. Instead, make one corner of the second tetrahedron, weave this into the first one, then lock the other three units into the second tetrahedron.

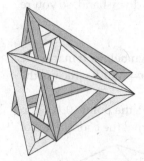

The first two tetrahedra make a sort-of 3D Star of David, with a corner of one tetrahedron poking through the side of the other, and a corner of the other poking through a side of the one. In fact, when the whole model is done **every** pair of tetrahedra should form such a 3D Star of David form.

The third tetrahedron is the most difficult one to weave into the model.

The figure to the right is drawn at a specific angle to help you do this. Notice how in the center of the picture there are three struts weaving together in a triangle pattern. If you look carefully, the same thing is happening on the opposite side of the model. As you insert your units for the third tetrahedron, try to form these triangular weaves and use them as a guide. In the finished model, there will be one of these triangular weave points under *every* tetrahedron corner.

These two types of symmetry—two tetrahedra making a 3D Star of David and the triangular weave points—are the best visual tools to use when inserting the units for the fourth and fifth tetrahedra. The pictures below also help.

What Is the Optimal Strut Width?

The instructions for Francis Ow's $60°$ unit have us start with 1×3 sized paper, which gives us a unit that is $1 \times 1/12$ in dimensions. In other words, if the side of one of the tetrahedra is 1, then the width of the strut in the tetrahedral frame that we make is $1/12$.

Is this the optimal strut width, or should we be using a wider or thinner strut for a more ideal fit? In this activity you'll use vector geometry and calculus to approximate the ideal strut width. This calculation is hard to do by hand, so you're better off using a computer algebra system to help.

The ideal strut width is given by the line segment L, shown below. It marks the shortest distance between the tetrahedron edge $\overline{v_3 v_4}$ and the point h which is the midpoint of the $\overline{v_1 v_2}$ edge from another tetrahedron.

We can use nice coordinates for v_1 and v_2 so that h will be the point $(0,0,1)$ on the z-axis. Then, since the tetrahedra fit inside a dodecahedron, the coordinates of v_3 and v_4 can be found to be as follows:

$$v_1 = (-1, 1, 1)$$
$$v_2 = (1, -1, 1)$$
$$v_3 = \left(0, \frac{-1+\sqrt{5}}{2}, \frac{1+\sqrt{5}}{2} \right)$$
$$v_4 = \left(\frac{1-\sqrt{5}}{2}, \frac{-1-\sqrt{5}}{2}, 0 \right)$$

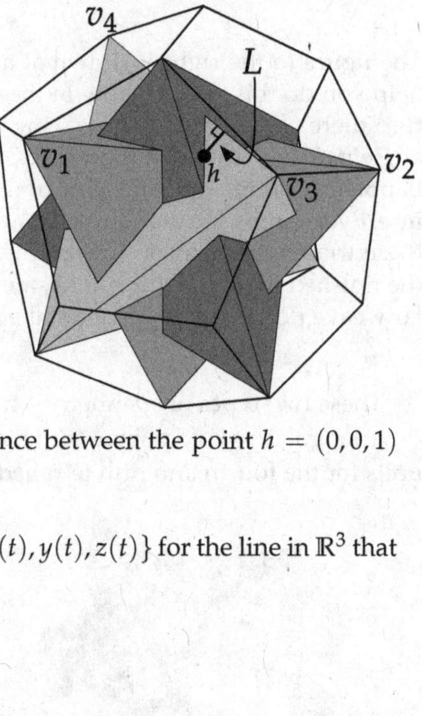

Our goal is to find $L =$ the minimum distance between the point $h = (0,0,1)$ and the line segment $\overline{v_3 v_4}$ (as shown above).

Question 1: Find a parameterization $F(t) = \{x(t), y(t), z(t)\}$ for the line in $\mathbb{R}^3$ that contains $\overline{v_3 v_4}$.

Question 2: Now find a formula for the distance between an arbitrary point $F(t)$ on the $\overline{v_3 v_4}$ line and the point $h = (0, 0, 1)$.

Question 3: Now minimize the distance function you found in Question 2 to find the length L. Hints: It might be easier to minimize the square of the distance function to get L^2.

Question 4: So what is the ideal strut width L? How does it compare to our use of struts that were 1/12 the side of a triangle?

COMMENTS, SOLUTION, AND PEDAGOGY

Comments

History of the model.   I conceived of making this model in 1995 while in graduate school at the University of Rhode Island. I had seen a Mathematica poster that depicted this object, but the width of the frames looked too narrow, as if the model would jangle in a loose tangle if it existed in reality. So I set out to make one via origami. I found Francis Ow's 60° unit ([Ow86]) to be perfect for this, especially since it can be made with any frame thickness desired. At the time I guessed, figuring that using 1×3 paper would work, giving edges of the frame that are $1 \times 1/12$ in dimensions. This is a little bit wider than the ideal, but it's close enough for a paper model. After I talked a crowd of my fellow graduate students into helping me fold the units, we collectively struggled to put it all together, and the model hung from the ceiling of the math department conference room for several years.

Once I saw that the model worked, I created instructions for it, posted them on my web site, and mailed a copy of them to Francis Ow, who lives in Singapore. He wrote back saying that he was surprised and delighted that such a complex model could be made from his unit. Since then this model has become very popular in origami circles and on the internet, even being voted onto the British Origami Society's list of "Top 10 Favorite Models" [Robi00].

Making the model.   Many people find the Five Intersecting Tetrahedra (FIT) model stunning to behold. Making one is *very* rewarding. It is up to instructors to decide how to go about teaching and making this model. Some instructors might want to build this object with their students for the first time, so that it'll be a discovery experience for everyone. Others might be more comfortable making an FIT themselves to become familiar with the process and the symmetries inherent in the model. If you do this, be sure to reserve plenty of time for your own study of this model. It is *not* easy to put together! Those who really want to challenge themselves should fold all 30 units, using five different colors, and try to assemble it using only a picture of the finished model (i.e., without the hints and figures on the handout). The truly masochistic can try it using only one color.

Actually, making yourself try it without the aid of the handout is a very good way to put yourself in the mindset of your students as they try to build this thing. It gets across how valuable understanding the symmetry of the finished object is when putting the model together.

Symmetries of the model.   When looking at the model, it's not hard to see that the corners of the tetrahedra would form a dodecahedron if we drew lines connecting nearby corners. There's a reason for this: It is possible to find four mutually equidistant corners in a dodecahedron. Thus, if we drew lines connecting these corners, we'd have a regular tetrahedron inscribed inside the dodecahedron. (See the illustration below.)

However, the dodecahedron has 20 corners, and this number is evenly divisible by four. This makes one suspect that you could inscribe five such tetrahedra inside a dodecahedron without using any corner more than once. This indeed works and gives us the solid known as the *compound of five tetrahedra*. (See below.)

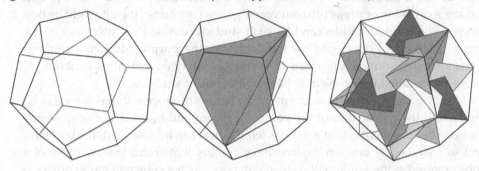

Thus, this model shares many symmetries with the dodecahedron. It has 120° rotational symmetry about the axes connecting two opposite tetrahedron corners (i.e., two opposite vertices of the dodecahedron), 72° rotational symmetry about the axes through points where five tetrahedra meet (i.e., through the midpoints of two opposite faces of the dodecahedron), and 180° rotational symmetry about the axes through the "midpoint" between two nearby tetrahedron corners (i.e., the midpoint of an edge of the dodecahedron). These rotational symmetries form a group, the rotational group of the dodecahedron, which is isomorphic to the alternating group A_5.

But the dodecahedron has reflection symmetries as well, and these are *not* symmetries of the FIT. This is because the FIT comes in two versions that are mirror reflections (a.k.a. enantiomorphic, a.k.a. chiral [Wei2]) of each other. (See below.) Indeed, if a whole classroom of students make their own FITs, then some of them will surely be enantiomorphic to each other. This can provide an opener for a class discussion about mirror symmetries in $\mathbb{R}^3$, which are often harder to visualize than those in $\mathbb{R}^2$.

Solution to the optimal strut width problem

The optimal strut width handout gives some very big hints. My thoughts on making this handout were centered around the fact that this is a very challenging problem. In fact, if instructors would like to use this problem as an advanced project for, say, a capstone or project-driven vector geometry course, then it might be best to keep this handout hidden and let such students devise their own way of doing it. Therefore, I developed the handout with the purpose of giving students a chance to see some applications of vector geometry and calculus material. In doing so, I wanted the problem to be doable for most students.

However, the handout does intentionally leave a number of things unexplained. First of all, the coordinates of the points v_1, v_2, v_3, and v_4 come out of nowhere. The idea behind them is that a regular tetrahedron can be inscribed inside a cube, and so why not let one tetrahedron have corners that coincide with those of a cube centered at the origin with side length two? Such a cube will have corners at $(\pm 1, \pm 1, \pm 1)$, and if we take the subset

$$(-1,1,1), (1,-1,1), (1,1,-1), (-1,-1,-1)$$

then they form the vertices of a regular tetrahedron. From this, the coordinates of the other four tetrahedra in the FIT can be found by rotating the first one about a suitable axis by multiples of $2\pi/5$. Details on how to actually do this are challenging and interesting, but they would also require a lot of class time, several more handouts, and a computer algebra system in order for students to have a chance of doing it. Interested readers should consult George Bell's excellent summary [Bell11] of the details of this approach.

Another important aspect that the handout glosses over is the rationale behind the point h. It is important for students to understand that the midpoint h between v_1 and v_2 is a place where we can measure the shortest distance to $\overline{v_3 v_4}$ to get the optimal strut width. And in order to see why this is so, students *must* have a finished FIT model to examine.

If one looks at a finished FIT model, one can see that at the midpoint of every tetrahedron edge there are two other tetrahedra struts touching. The below figure illustrates this, where the black dot shows the midpoint of one of the tetrahedron edges.

In other words, the point h is a spot where the bottom of the strut of one tetrahedron is touching the top edge of another, and therefore the optimal strut width can be determined from this point by finding the distance between this point and the line made by the tetrahedron edge.

All this might seem pretty questionable to a reader of this book, but all we are doing is capitalizing on yet another symmetry of the FIT. It is a kind of symmetry that is very apparent when examining a physical FIT model in one's hands, and so students will understand the reasoning behind the point h after some prodding and scrutinizing of the model.

Question 1. Written as a vector function, the two lines can be most easily expressed by

$$F(t) = (v_3 - v_4)t + v_4.$$

Simplifying in Mathematica, this gives us

$$F(t) = (v_3 - v_4)t + v_4$$
$$= \left(\frac{-1+\sqrt{5}}{2}t + \frac{1-\sqrt{5}}{2}, \sqrt{5}t - \frac{1+\sqrt{5}}{2}, \frac{1+\sqrt{5}}{2} \right)$$

Question 2. It is easier to use the square of the distance between an arbitrary point $F(t)$ on $\overline{v_3 v_4}$ and $h = (0,0,1)$. Computationally, this can be expressed with a dot product:

$$H(t) = (\mathrm{dist}(F(t),h))^2 = (F(t) - h) \cdot (F(t) - h).$$

This, of course, is just mimicking the standard distance formula. Luckily, it simplifies to something reasonable:

$$H(t) = 8t^2 - (9 + \sqrt{5})t + 4.$$

Question 3. The function $H(t)$ is quadratic in t and concave up, so its single critical point will be the minimum that we seek. We get that $H'(t) = 16t - (9 + \sqrt{5})$, and so the t value that will give us the minimum squared distance between $\overline{v_3 v_4}$ and h is

$$t_0 = \frac{9 + \sqrt{5}}{16}.$$

To turn this into the minimum distance, we need to plug this value back into $H(t)$ and take its square root. This arduous task can be done by hand, but it is more easily handled with a computer algebra system like Mathematica or Maple. After simplifying, we obtain

$$H(t_0) = \frac{21 - 9\sqrt{5}}{16} \quad \text{and} \quad \sqrt{H(t_0)} = \frac{1}{4}\sqrt{21 - 9\sqrt{5}}.$$

Thus we get that the optimal strut width is $L = \sqrt{21 - 9\sqrt{5}}/4 \approx .2339$.

Question 4. In order to have the answer to Question 3 make sense, we need to look at the ratio between the length L that we found and the length of an edge the tetrahedra. This is because the origami units of the FIT span an entire tetrahedral edge, and so the ratio of L over the tetrahedron edge length will give us something to compare to the $1/12$ ratio of the origami units we've been folding.

The easiest way to find the tetrahedral edge length is to compute

$$\text{dist}(v_1, v_2) = 2\sqrt{2}.$$

Thus the ratio R that we are looking for is

$$R = \frac{\sqrt{21 - 9\sqrt{5}/4}}{2\sqrt{2}} = \frac{1}{8\sqrt{2}}\sqrt{21 - 9\sqrt{5}}.$$

By using square root manipulations that are too tedious to reproduce here, we can obtain

$$R = \frac{\sqrt{3}}{12 + 4\sqrt{5}} = \frac{\sqrt{3}}{8\varphi^2},$$

where $\varphi = (1 + \sqrt{5})/2$ is the golden ratio.

Converting this to a decimal, we get $R = 0.0826981...$, whereas our origami units have ratio $1/12 \approx 0.8333$. Interesting! Our Francis Ow units, folded from 1×3 paper, turn out to give us tetrahedral struts that are within 0.000635 of the optimal width. That is 3 decimals of accuracy, which is probably too small to really notice with an origami model.

Perhaps a more concrete way to think about it would be to assume that we are starting with 1×3 paper that is 10 inches on the long side. Then our Ow $60°$ units will make a tetrahedral frame with strut width approximately 0.83333 inches. An ideal width would be about 0.82698 inches, so we're off by only 0.00635 of an inch or approximately $.16$ of a millimeter. That is hardly noticeable at all!

In other words, using 1×3 paper really is fine for origami purposes because paper is flexible, so the slightly thicker origami struts won't matter. If, however, we were to make a version of the FIT structure out of wood or glass, as wood-working master Lee Krasnow and glass artisan Hans Schepker have done, then the question of accuracy becomes very important. Does a fraction of a millimeter of error matter if making the model out of wood or glass? Or what if the size of the finished piece is to be a lot larger than one made from 10 inch-long rectangles? An artist working with rigid materials might prefer to err on the side of thinner strut widths than the optimal.

Other methods. There are other ways to approach this problem. Don Barkauskas (University of Arizona) suggests a way that relies only on vector methods. Use the cross product to find the unique direction vector v mutually perpendicular to the two lines $\overline{v_1v_2}$ and $\overline{v_3v_4}$. Then, find the equation of the plane containing v and $\overline{v_1v_2}$ and the equation of the plane containing v and $\overline{v_3v_4}$. The intersection of these two planes will form a line M. The length of the line segment between the point where M intersects $\overline{v_1 v_2}$ and the point where M intersects $\overline{v_3 v_4}$ will be the minimum distance between the two tetrahedron edges. This is not the optimal distance L, however; it is the height of the triangular cross section of the optimal strut, and either trigonometry or other simple triangle geometry must then be used to find the desired length L. While this approach avoids calculus, it does require more involved steps, like finding the dihedral angle of an edge of a regular tetrahedron (which is *not* 60°) that is needed for the cross section of the strut.

Another way to avoid calculus, proposed by Kyle Calderhead (Illinois College), is to also determine the vector mutually perpendicular to $\overline{v_1 v_2}$ and $\overline{v_3 v_4}$, except this time make it a unit vector. Call this vector v_u. Then, take the dot product of v_u and a vector w pointing from a point on $\overline{v_1 v_2}$ to a point on $\overline{v_3 v_4}$. This dot product will give the length of w projected onto v_u, which will be the minimum distance between the two lines $\overline{v_1 v_2}$ and $\overline{v_3 v_4}$. This still, however, requires one to convert this minimum distance into the distance L that we want, which requires the cross section triangle of a strut and the dihedral angle of a tetrahedron.

Pedagogy

As mentioned previously, instructors should practice making the FIT model themselves, perhaps even experimenting with different sizes and weights of paper, before expecting students to do it. Otherwise, it is entirely up to you how to incorporate this into various classes. A math for liberal arts class will find this model very challenging, and it should be viewed as a difficult puzzle that also showcases some complex polyhedral structures and symmetries. Depending on the skills of your students, it might be better to only expect them to make one tetrahedral frame in class, and then unveil your own model of the full FIT. Motivated students can then take it upon themselves to make one of their own, perhaps for extra credit.

Students in an undergraduate geometry or multivariable calculus course should be able to do the entire activity. Structuring the activity is entirely up to you. Some people have had much success by forming students into groups of three or four, trying to make sure that each group has at least one good folder and one person with solid math/visualization skills. When Kyle Calderhead used this approach, he commented, "In most groups, the ace folder was not the ace mathematician, so it seemed like most everyone felt like they had something to contribute."

Activity 17
MAKING ORIGAMI BUCKYBALLS

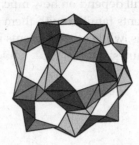

For courses: geometry, graph theory, topology

Summary

This activity has multiple parts.

(a) Students learn the PHiZZ unit and use 30 units to make a dodecahedron with either a proper 3-edge-coloring or a symmetric 5-edge-coloring.

(b) Students find a Hamilton circuit on the graph of the soccer ball (C_{60} Buckyball, truncated icosahedron) and use it to plan a proper 3-edge-coloring. Then, they (perhaps working in teams) make a 90-unit PHiZZ version.

(c) Students use Euler's formula and counting tricks to prove that every Buckyball has exactly 12 pentagons. A much bigger project is to classify all spherical Buckyballs and develop a formula for the number of PHiZZ units needed to make them.

Content

In a graph theory course PHiZZ units can be a way to give students hands-on experience with 3-edge-colorings. Hamilton circuits, edge colorings, Euler's formula, and counting techniques are standard topics in undergraduate graph theory courses, and students are usually very eager to (either individually or by working together) make large Buckyballs. Coxeter has a nice classification of spherical Buckyballs that does not seem to be very well-known, which offers a very nice way to show how subjects like graph theory, combinatorics, polyhedra, and vector geometry can be tied together. This material can easily take up a week or more of a graph theory course, but instructors can decide how much or how little they want to do. Also, this activity uses a lot of standard material, so it might be worthwhile to spend time on PHiZZ units as a way to introduce several concepts.

Handouts

There are three handouts relating to the three parts of this activity.

Time commitment

For the first handout, students can fold 3–5 units and learn how to lock them together in 20 minutes. Folding all 30 units might be best done outside of class. The speed of the second handout will depend on how much experience your students have with planar graphs; students familiar with them will take only 15 minutes, working in groups, to finish this, while other students may need 30–40. The third handout will take about 30 minutes.

The PHiZZ Unit

This modular origami unit (created by Tom Hull in 1993) can make a large number of different polyhedra. The name stands for **Pentagon Hexagon Zig-Zag** unit. It is especially good for making large objects, since the locking mechanism is strong.

Making a unit: The first step is to fold the square into a 1/4 zig-zag.

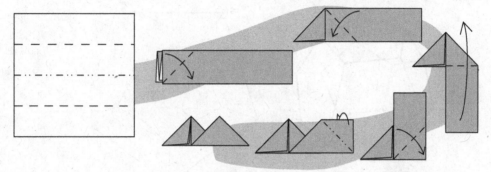

When making these units, it's important to make all your units **exactly** the same. It's possible to do the second step backwards and thus make a unit that's a **mirror image** and won't fit into the others. Beware!

Locking them together: In these pictures, we're looking at the unit "from above." The first one has been "opened" a little so that the other unit can be slid inside.

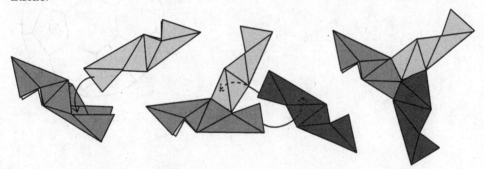

Be sure to insert one unit **in-between** layers of paper of the other. Also, make sure that the flap of the "inserted" unit hooks over a crease of the "opened" unit. That forms the lock.

Assignment: Make **30 units** and put them together to form a **dodecahedron** (shown to the right), which has all pentagon faces. **Also** use only 3 colors (10 sheets of each color) and try to have no two units of the same color touching.

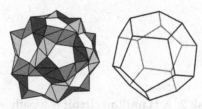

Planar Graphs and Coloring

Drawing the **planar graph** of the polyhedron can be a great way to plan a coloring when using PHiZZ units. To make the planar graph of a polyhedron, imagine putting it on a table, stretching the top, and pushing it down onto the tabletop so that none of the edges cross. Below is shown the dodecahedron and its planar graph.

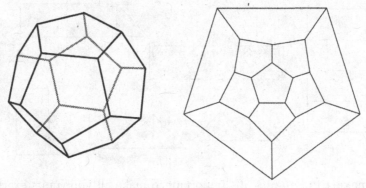

Task 1: Draw the planar graph of a soccer ball. Make sure it has 12 pentagons and 20 hexagons.

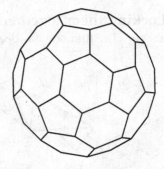

Task 2: A **Hamilton circuit** is a path in a graph that starts at a vertex, visits every other vertex, and comes back to where it started without visiting the same vertex twice. Find a Hamilton circuit in the planar graph of the dodecahedron.

When making objects using PHiZZ units, it's always a puzzle to try to make it using only 3 colors of paper with no two units of the same color touching. Each unit corresponds to an edge of the planar graph, so this is equivalent to a **proper 3-edge-coloring** of the graph.

Question: How could we use our Hamilton circuit in the graph of the dodecahedron to get a proper 3-edge-coloring of the dodecahedron?

Task 3: Find a Hamilton circuit in your planar graph of the soccer ball and use it to plan a proper 3-edge-coloring of a PHiZZ unit soccer ball. (It requires 90 units. Feel free to do this in teams!)

Making PHiZZ Buckyballs

Buckyballs are polyhedra with the following two properties:

(a) each vertex has degree 3 (3 edges coming out of it), and

(b) they have only pentagon and hexagon faces.

The PHiZZ unit is great for making Buckyballs because you can make pentagon and hexagon rings:

These represent the faces of the Buckyball. But when making these things, it helps to know how many pentagons and hexagons we'll need!

To the right are shown three Buckyballs: The dodecahedron (12 pentagons, no hexagons), the soccer ball (12 pentagons, 20 hexagons), and a different one. (Can you see why?)

Question 1: How many vertices and edges does the dodecahedron have? How about the soccer ball? Find a formula relating the number of vertices V and the number of edges E of a Buckyball.

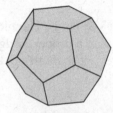

Question 2: Let F_5 = the number of pentagon faces in a given Buckyball. Let F_6 = the number of hexagon faces. Find formulas relating
(a) F_5, F_6, and F (the total number of faces). (Easy!)

(b) F_5, F_6, and E. (Harder.)

Question 3: Now use Euler's formula for polyhedra, $V - E + F = 2$, together with your answers to Questions 1 and 2, to find a formula relating F_5 and F_6, the number of pentagons and hexagons.

Question 4: What can you conclude about all Buckyballs?

SOLUTION AND PEDAGOGY

Handout 1: The PHiZZ Unit

I invented the PHiZZ unit in 1993 while in graduate school. My aim was to design a unit that had a strong enough locking mechanism to support the construction of very large polyhedra. The result worked—a full 1/4 of the paper is devoted to each lock, and it had the added bonus of forming "rings," which made it a lot easier to see the faces of the underlying polyhedral structure. However, I felt that the unit did not support making triangle and square rings, since these forced the paper to buckle and, when certain types of origami paper were used, fall apart. Thus I had to restrict myself to only pentagon and hexagon faces, which created the name Pentagon-Hexagon Zig-Zag Unit (or PHiZZ for short). I later discovered that heptagon and larger faces could be made, but that these would introduce negative curvature. See the Making Origami Tori activity for information on how to incorporate this into models.

Since the heart of this activity is centered around folding PHiZZ units and putting them together, instructors should spend a substantial amount of time beforehand making and playing with PHiZZ units themselves. Some people, faculty and students alike, find the locking mechanism difficult to comprehend from the diagrams on the handout. Be sure to pay *close* attention to the drawings and their depiction of how the flaps of one unit are to be inserted between the layers of another. At the very least make 30 units to form the dodecahedron, and use the planar graph to get a 3-edge-coloring. Better preparation would be to fold 90 units to make the soccer ball (a.k.a. Buckminster fullerene, a.k.a. truncated icosahedron), which really is quite an impressive model to behold. Follow the handout to use a Hamilton circuit to generate a proper 3-edge-coloring. Such models make great decorations to hang in one's office, by the way!

I find that the ideal paper to use for these units is "memo cube paper" that can be found in office supply stores. Make sure to avoid Post-It notes, though, as the sticky strip will get in the way of the unit's functionality. If you can find it, buy memo cube paper that comes in its own plastic box/holder. Such paper is much more accurately square than other memo cube paper. (And non-square paper can be slightly problematic in making accurate units.)

Normal origami paper is useful as well, although it needs to be cut down to smaller squares. For example, when making very large Buckyballs, say with 500 or more units, using 3 inch memo cube paper might result in a model too large for one's dorm room. Instead try cutting normal origami paper (the kind that is colored on one side and white on the other) into 2 inch or 2.5 inch squares. This tends to be much more manageable.

Accuracy in the units does help, and some effort will have to be made in class to make sure that the students' units are decent. They should not look like they were folded by someone wearing mittens.

But more importantly, notice that the units can come in *left-* and *right-handed* versions. If you follow the instructions carefully, all your units will be right-handed and will lock together properly. But once you get the hang of the folds and start making them without looking at the instructions, it can be easy to accidentally make a unit that is a mirror-image of the others (i.e., left-handed). Such a unit will *not* be able to lock with other, opposite-handed units. So make sure your students are aware of this pitfall!

Once your class folds a few units and learns how to lock them, you may find your students making piles of vertices—three units locked to form a pyramidal vertex—hoping to then join them together to make the dodecahedron. This is a *bad* approach. It is very difficult to join three vertex clusters together to make a new vertex in-between them. Anyone who tries this will become frustrated and have to take their vertices apart. The best way to make things out of PHiZZ units is to form one vertex and then keep adding onto it with more units, building your polyhedron one vertex at a time. Suggesting this to students can eliminate a lot of headaches.

Handout 2: Planar Graphs and Coloring

The first task here is to draw the planar graph of a soccer ball, otherwise known as the truncated icosahedron. Students usually enjoy these kinds of activities a lot, but often they need some help on how to do them. Demonstrating how the planar graph of the dodecahedron can be made can help; start by drawing a pentagon, then notice that pentagons must be drawn around it, and each vertex must have degree 3, etc. For the soccer ball, also start with a pentagon face, and draw hexagons adjacent to every side of the pentagon, and work your way out. Encourage students to make their drawings as symmetric as possible, as below, for example.

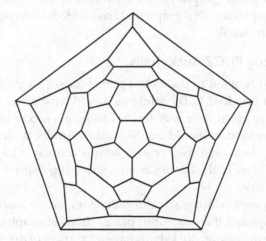

Then, students are asked to consider Hamilton circuits on these graphs. The reason for this is because Hamilton circuits can provide an easy way to generate a proper 3-edge-coloring on the graphs. Here's how: Once you have a Hamilton circuit, color the edges on the circuit with two colors, alternating as you go along. One can prove that on any cubic (all vertices degree 3) graph, we must have an even number of vertices. (See the third handout, Question 1 solution for a proof.) Since the Hamilton circuit visits every vertex exactly once, this means our Hamilton circuit will have an even number of edges, and thus we will be able to 2-color the circuit properly. Then we can color all the remaining, non-circuit edges with the third color, and bingo! We have our proper 3-edge-coloring.

There are many different ways to find a Hamilton circuit on the dodecahedron and the soccer ball. Below are shown one example of each.

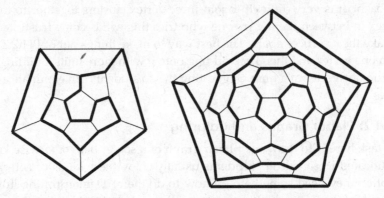

There is a lot of more interesting graph theory to explore here. For example, back in the 1890s Tait tried to use the concept of Hamilton cycles to prove the Four Color Theorem (which was then still a conjecture). This is done through an elegant method (due to Tait) of transforming a 4-face-coloring of a planar graph into a 3-edge-coloring of a cubic planar graph. Tait's mistake, however, was in assuming that all 2-connected planar graphs have a Hamilton circuit. Indeed, in the 1930s Tutte found an example of such a graph that has no Hamilton circuit. See [Bar84] and [Bon76] for more details.

Handout 3: Making PHiZZ Buckyballs

This handout would be good to use a class period (or so) after first encountering the PHiZZ unit and Handout 1. The students should have made a PHiZZ dodecahedron and perhaps be on their way toward making a soccer ball. Envisioning larger and larger Buckyballs should not be hard for students. Just remind them that every geodesic dome that they've ever seen is actually a large Buckyball of some sort. Find a picture of the Epcot Center's Spaceship Earth if you really want to drive the point home.

(Actually, most geodesic dome structures that you see are *duals* of a Buckyball. If your class has explored the concept of planar duals of graphs, this will be an interesting example to explore; Buckyballs have all vertices of degree 3, while their duals, geodesic spheres, have all triangle faces. Buckyballs have only pentagon and hexagon faces, while geodesic spheres have only vertices of degree 5 and 6.)

The three Buckyballs shown on the handout are the dodecahedron (a "trivial" Buckyball), the soccer ball (which is the classic carbon-60 molecule, which chemists often call a Buckminster Fullerene), and a third, bigger Buckyball that students will be unfamiliar with. This third one is fundamentally different from the soccer ball because, as you can see, it has vertices where three hexagons meet; all vertices of the soccer ball have a pentagon meeting two hexagons. Inquisitive students, and perhaps you yourself, will find this puzzling—if three regular hexagons meet, we get a flat plane with no curvature. So how could this make a polyhedron? This reasoning is correct, and it proves that the hexagons in this object are *not regular*. In order for this arrangement of pentagons and hexagons to form a polyhedron, the hexagons need to be a bit irregular. (This is why the image on the handout, generated using Mathematica, looks a little odd.) Luckily, the PHiZZ unit is flexible enough to make such hexagons slightly irregular, so if you or your students try making this Buckyball, you won't notice the difference at all.

Question 1. After playing with the PHiZZ unit for a while, students will have everything that they need to figure out how many vertices and edges a dodecahedron and the soccer ball have. Make the students count these things themselves, and be firm about this! The whole point of having the students construct origami polyhedra is for the hands-on experience to build conceptual understanding of the objects that they build. Asking these kinds of questions brings such concepts to the forefront, but students need to discuss and wrestle with the questions themselves to get it.

In any case,

	vertices	edges
dodecahedron	20	30
soccer ball	60	90

This suggests the equation $V = 2E/3$. But this formula can be proven for Buckyballs in general: Imagine that we take any Buckyball and visit each vertex, counting the number of edges coming out of that vertex. Of course, we'll count three edges at each vertex, counting a total of $3V$ edges. But each edge will have been counted twice! This is because each edge connects two vertices, so our visits to each of those vertices will have counted that edge. Thus we have $3V = 2E$.

Notice that this immediately proves that every Buckyball has an even number of vertices (or any 3-regular graph, for that matter).

I want to emphasize how useful this type of counting argument is for studying the combinatorics of polyhedra. In fact, we'll be using it again in the next question.

Question 2. In part (a) all I'm looking for is $F = F_5 + F_6$. Yes, it's that easy.

Part (b) requires a counting argument similar to the one in Question 1, except that this time we'll visit each face of the Buckyball. We're still counting edges, and this time we count the edges that surround each face that we visit. All the pentagon faces will give us 5 edges, and so we'll count $5F_5$ edges from the pentagon faces. From the hexagons we will count $6F_6$ edges. And once again we will have counted each edge twice (since each edge borders two faces)! Thus,

$$5F_5 + 6F_6 = 2E.$$

Question 3. All the equations that we have should do something for us here, and there are several ways to get the desired result. Following the lead with Euler's formula, let's use $V = 2E/3$ to eliminate the V variable:

$$F - \frac{1}{3}E = 2.$$

Now, we want a formula involving F_5 and F_6, so let's use $F = F_5 + F_6$ and $2E = 5F_5 + 6F_6$ to obtain an equation with only these two variables:

$$F_5 + F_6 - \frac{1}{3}\left(\frac{5F_5 + 6F_6}{2}\right) = 2$$

$$\Rightarrow 6F_5 + 6F_6 - 5F_5 - 6F_6 = 12$$

$$\Rightarrow F_5 = 12.$$

Wow! The number of hexagons just dropped out and gave us a fixed number of pentagons! So Question 3 is sort of a "trick" question, in that the formula involving F_5 and F_6 doesn't contain F_6 at all.

But this does make **Question 4**'s answer clear: Every Buckyball has exactly 12 pentagon faces, no more, no less.

Follow-up ideas

That $F_5 = 12$ always is pretty surprising, and it can lead into much more extensive studies of Buckyball and geodesic sphere structures. Other Buckyballs can be made by drawing planar graphs with all vertices of degree 3, 12 pentagon faces, and some number of hexagons. For example, you can challenge students to come up with as many cubic graphs with 12 pentagon faces and only two hexagon faces as possible. (These can then be made using how many PHiZZ units?) Is it possible to have only one hexagon face in such a graph? (The answer is, "No!")

Other facts can be discovered by examining such models. Beta-tester Jason Ribando of the University of Northern Iowa notes, "It may be worth noting in the instructor's notes that the pentagon holes on parallel faces of the PHiZZ dodecahedron are aligned, unlike the Platonic solid version. It could make for a good exercise to explain why!" 1993 HCSSiM student Gowri Ramachandran noticed that when we properly 3-edge-color the dodecahedron, faces on opposite sides of the polyhedron will have similar colorings (i.e., if one face has, say, two yellow edges, two pink edges, and one white edge, then so will the opposite face). Does this persist in larger spherical Buckyballs?

Here is another question: When we 3-edge-color a PHiZZ Unit dodecahedron using colors red, blue, and green, some of the vertices will have the colors arranged in this order (red, blue, green) going clockwise about the vertex and others will have this order going counterclockwise. How many are clockwise and how many are counterclockwise? Is this always true? And are these arranged in any kind of pattern on the dodecahedron? Clearly there are many questions to explore in coloring spherical Buckyballs, making this especially fertile ground for student research.

Students interested in chemistry might like making PHiZZ unit objects that model what nanotechnology scientists are exploring. For example, consider the images found on Richard E. Smalley's web page at Rice University, http://smalley. rice.edu/smalley.cfm?doc_id=4866. Smalley was one of the people who won a Nobel Prize for their work on discovering the carbon-60 (Buckminster Fullerene) molecule, and his newer research in Bucky tubes may end up revolutionizing superconductivity.

Geodesic dome structures are spherical, however. To make a Buckyball as spherical as possible, we need to think of the 12 pentagons as being evenly distributed with hexagons in between them. In fact, we can think of each pentagon as corresponding to a vertex of the icosahedron, and each triangle face of the icosahedron will represent three pentagons and the hexagons nested in between them on the spherical Buckyball. (See the picture below.) These triangle "tiles" can uniquely determine and enumerate spherical Buckyballs as well as explain their symmetry group [Hull05-2].

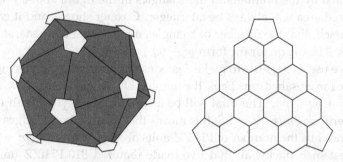

Coxeter [Coxe71] presents a classification of such Buckyball duals using triangle tiles on the triangular lattice, and this work leads to explicit formulas for the number of vertices, edges, and faces of any spherical Buckyball. To give a brief summary, the idea is to consider the dual of such tiles, which would give a triangular "tile" of a geodesic sphere. These can be completely classified by taking three mutually equidistant points on the triangular lattice. That is, consider the lattice formed by integer linear combinations of the vectors $v_1 = (1, 0)$ and $v_2 = (1/2, \sqrt{3}/2)$. The integer multiples of v_1 will form the p-axis of this lattice and integer multiples of v_2 will form the q-axis. Let one of the corners of our triangle tile be $(0, 0)$ and let another be an arbitrary point (p, q) on the lattice. This will determine the third point needed to make the tile, which can be found by rotating (p, q) about the origin by $60°$. An example in which we take $(p, q) = (2, 1)$ (which is really the point $2v_1 + v_2$ on the Cartesian plane) is shown below.

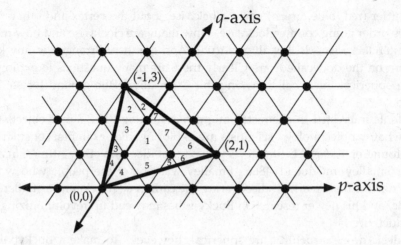

The nice thing about this approach is that we can compute the area of these triangle tiles. If we normalize this area so that the area of one triangle on the lattice equals one, then we only need to count how many unit triangles are in the tile to calculate its area. The tile's symmetry will guarantee that any triangles on the edge of the tile that are cut-off will have a matching pair somewhere else. (This is demonstrated by the numbers in the triangles in the figure above.) Therefore, this normalized area will always be an integer. Coxeter shows, and it can be fun to prove yourself, that the number of triangles in a triangle tile generated by the point (p, q) will be the quadratic form $p^2 + pq + q^2$.

Thus, if we use a (p, q)-tile to make a geodesic sphere, we'll be placing one tile on each face of an icosahedron. Thus, the number of triangle faces on such a sphere will be $20(p^2 + pq + q^2)$. The dual will be a spherical Buckyball with this same number of vertices. Since $3V = 2E$, this means that the number of edges in such a Buckyball, and thus the number of PHiZZ units needed, will be $30(p^2 + pq + q^2)$.

The largest such Buckyball that I've made required 810 PHiZZ units, made from a (3,3)-tile. Pictures can be found at http://mars.wne.edu/~thull/gallery/modgallery.html.

Activity 18
MAKING ORIGAMI TORI

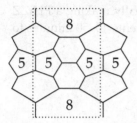 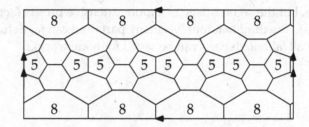

For courses: geometry, graph theory, topology

Summary

Students who have already made some PHiZZ unit models (at least the dodec-ahedron) are asked to try making a PHiZZ unit torus. This leads to discussions of positive versus negative curvature and the fundamental domain of a torus. To help plan their torus designs, the combinatorics of "Bucky tori" are studied.

Content

This can make a great introduction to the topology of toroidal surfaces. It also offers a chance for graph theory students to get their hands dirty with graphs of surfaces other than the plane. The combinatorial study uses Euler's formula for the torus, $V - E + F = 0$, to prove things such as every three-valent toroidal graph with only pentagon, hexagon, and heptagon faces must have an equal number of pentagons as heptagons.

   This is a continuation of the Making Origami Buckyballs activity, although all it really requires is the first handout from that activity.

Handouts

There are three handouts:

(1) Explores making bigger PHiZZ unit rings (with negative curvature).

(2) Explores drawing toroidal graphs on a fundamental domain.

(3) Explores Euler's formula on orientable surfaces of genus g. This leads to finding relationships between the number of pentagons and the number of n-gons in "Bucky tori."

Time commitment

The only time sink for the first handout is in folding the units needed to make the rings. If these are made in advance, this will only take 15–20 minutes. The second handout only involves drawing toroidal graphs. It takes 10–15 minutes for the first page, but the second page is a more extensive project (actually making a PHiZZ torus). The third handout has many parts and can take students 40–50 minutes to do it all (although parts can be saved for homework).

Bigger PHiZZ Unit Rings

This handout asks you to experiment with making larger "rings" using PHiZZ units.

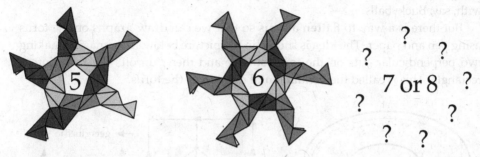

Activity: Make a heptagon or octagon ring out of PHiZZ units (it'll require 14 or 16 units, so feel free to do it in groups). This will be challenging: How can you make the ring close up? Do not force any extra creases in the units! They should go together just like normal.

Question: Compare what a pentagon ring, a hexagon ring, and a bigger ring (like a heptagon or octagon ring) look like.

Specifically, imagine these rings lying on a surface. What kind of surface would the pentagon ring be lying on?

How about a hexagon ring?

How about a heptagon or octagon ring?

So, if you were to make a **torus** (i.e., a doughnut) using PHiZZ units, where on the torus might you place your pentagons, your hexagons, and your bigger-gons?

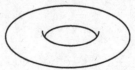

Drawing Toroidal Graphs

When planning a PHiZZ unit torus model, it can be hard to visualize what you're doing because you can't just draw the planar graph of the structure like you can with, say, Buckyballs.

But there is a way to **flatten** a torus so that we can draw graphs on the torus using pen and paper. The idea is shown in the picture below. You imagine making two perpendicular cuts on the torus surface and then "unroll" the torus into a rectangle. This is called the **fundamental domain of the torus**.

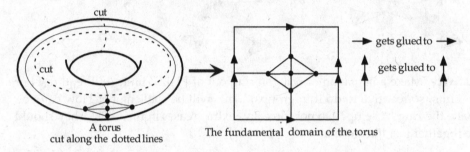

A torus
cut along the dotted lines

The fundamental domain of the torus

gets glued to
gets glued to

The idea in the fundamental domain is that any edge you draw that hits the boundary must come back on the other side. Thus a graph drawn on the torus, like the one shown above, can be represented on the fundamental domain by making some edges "wrap around" from top to bottom and from left to right.

Activity: Draw the graph of the **square torus** (shown below right) on a fundamental domain.

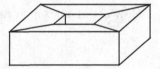

You now have what you need to start designing your own PHiZZ unit torus. Just start with the fundamental domain of a torus and try to draw a graph on it that has

(1) all vertices of degree 3 and

(2) only pentagon, hexagon, or higher faces.

(Square and triangle faces don't work very well with the PHiZZ unit.)

Unfortunately, making PHiZZ unit tori can take a lot of units. People have made ones using hundreds of units. But, they can be made with a more reasonable number. Below is a torus, designed by mathematician sarah-marie belcastro, that requires 84 units. It's made from a small pattern (below left, in the dotted box) that is repeated four times on the fundamental domain (below right). It uses only pentagon, hexagon, and octagon faces.

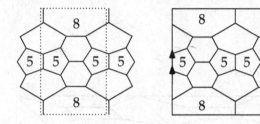

You can make the above torus or try designing your own. You might be able to design a smaller one by using larger polygons, like 10-gons, instead of octagons.

Advice: When making such a torus, make the larger, negative curvature polygons on the inside rim **first**. This may seem hard, but it's a lot easier to do them at the beginning than waiting until the end. Once you have the inner rim in place, it's a lot easier to then make the hexagons and pentagons.

Euler's Formula on the Torus

Question 1: Below is shown a **square torus**. What does Euler's Formula, $V - E + F$, give for this polyhedron?

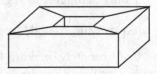

Question 2: How about for a **2-holed torus**?

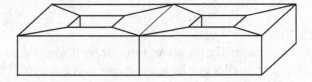

Question 3: We define the **genus** of a polyhedron to be the number of "holes" it has. (So a torus has genus 1, a two-holed torus has genus 2, an icosahedron has genus 0, etc.) Find a **generalized Euler's formula** for a polyhedron with genus g.

Properties of Toroidal "Buckyballs"

Now that you know Euler's Formula for the torus, we can learn some things that will help you plan making tori using PHiZZ units.

Question 4: Suppose that we make a torus using PHiZZ units and only making **pentagon**, **hexagon**, and **heptagon** (7-sided) faces. Find a formula relating F_5 (the number of pentagon faces) and F_7 (the number of heptagon faces).

Hint: Remember that we still have $3V = 2E$. Use the same techniques that we used to prove that all Buckyballs have only 12 pentagon faces.

Question 5: Suppose that we made a PHiZZ unit torus using only **pentagon**, **hexagon**, and **octagon** faces. Find a formula relating the number of pentagon and octagon faces.

Question 6: Can you generalize these results?

SOLUTION AND PEDAGOGY

This activity really needs to be preceded by the Making Origami Buckyballs activity. For one thing, that's where you'll find instructions for the PHiZZ unit to pass out, and no one can be expected to make origami tori from PHiZZ units without exploring spherical models (like Buckyballs) first. Also, many of the counting arguments used in the torus activities (especially in the Euler's Formula on the Torus handout) are similar to those developed in the Buckyball activities.

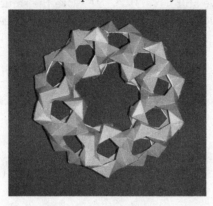

As with the Buckyball activity, instructors will have to spend some time making and experimenting with PHiZZ unit tori on their own. This can involve a significant time commitment, as even small PHiZZ tori come close to 100 units. (The torus in the photograph above is made from 105 units.) I recommend making belcastro's example in the second handout that requires 84 units. However, you can make this smaller; the example in the handout uses four copies of the basic structure. You can instead use only three, and this will require only 63 units. But this version is much harder to put together, since this puts a lot more tension on the units.

Below is a PHiZZ torus of my own design. The basic structure uses decagons and three copies of itself to complete the torus. It requires 81 PHiZZ units.

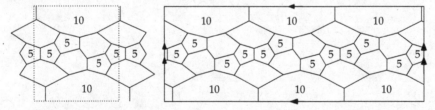

One tip when making these is to build the ring of large "polygons" first, which will be the inside part of the torus surface. This is where all the negative curvature comes into play, and it's the hardest part to conceptualize and execute. Once that is put together, adding the hexagon and pentagon polygons will be much easier. (And as with the Buckyballs, putting in the last units is always tricky, but it's a *lot*

easier to do as part of a pentagon or hexagon on the outside rim of the torus than on the inside.)

Making PHiZZ unit tori can be immensely satisfying. Students may, perhaps, have made cardboard polyhedra before, which can mirror the construction methods used in making modular origami polyhedra. But having a chance to construct an actual torus is much more rare, and this activity offers an opportunity for students to develop substantial intuition about the types of polygons and curvature elements that must come together to make a torus.

Handout 1: Bigger PHiZZ Unit Rings

This activity is only meant to get the students to explore what took me many years of playing with PHiZZ units to discover: You can make rings larger than hexagons, but they induce negative curvature! When trying to make such rings for the first time, it may seem impossible. Once you get enough units in place for a hexagonal ring, it doesn't seem like any more will fit. But if you allow the ring to twist in space, more sides can be inserted, giving us negative curvature.

Thus the answers to the questions asked on the handout are

- a pentagon ring could lie on the surface of a dome or sphere,

- hexagon rings are flat, and thus can lie on a flat plane,

- heptagon and larger-gons can lie on a saddle point surface, like a hyperbolic paraboloid or Pringles™ chip surface, and

- the pentagons will have to go on the outer part of the torus, where positive curvature exists. The heptagon, octagon, or larger rings will need to be on the inner part, where there's negative curvature.

Handout 2: Drawing Toroidal Graphs

This handout introduces the concept of the fundamental domain of a torus as used for the purposes of drawing toroidal graphs. If students have already seen this concept, then all the better. The basic nature of the first page of this handout would make it suitable in, say, a rubber-sheet topology course soon after the fundamental domain concept was introduced. This is because actually drawing toroidal graphs on a fundamental domain is a great way to solidify the concept.

The square torus graph is shown below.

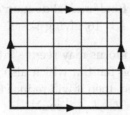

The remainder of this handout asks students to try making "Bucky tori" out of PHiZZ units. Following the belcastro design given is challenging, and since it requires a large number of units, students could be encouraged to work together to fold the units. I should emphasize, however, that this modular origami project is difficult. Students who haven't spent much time making spherical PHiZZ unit structures will find making PHiZZ tori nearly impossible. On the other hand, making these tori can be so satisfying and educational that struggling through the tricky mechanics of putting it together can be very worthwhile.

An easier (perhaps) project is for students to merely design toroidal graphs that could be made using PHiZZ units, like the belcastro example given. Giving constructive guidance to students for this kind of task takes some practice. I've found it best to let the students try it on their own for a while and then check how they're doing, making sure that all their vertices have degree 3, all their faces are pentagons or larger, and all edges that cross the boundary of the fundamental domain enter again at the appropriate spot.

It's very easy to design toroidal graphs with only pentagon, hexagon, and higher faces that are too small to be made from PHiZZ units, but experimentation on making smaller and smaller PHiZZ tori should be encouraged. At the 2000 Hampshire College Summer Studies in Mathematics, sarah-marie belcastro and I challenged our students to do just this, and the resulting "Torus Wars" resulted in some excellent designs requiring less than 100 units.

Of course, very large tori can be designed as well. The largest that I've made uses 660 PHiZZ units and was on display for several years at the Origamido Studio in Haverhill, MA. As with large Buckyballs, constructing such large objects can be the goal of student group projects.

Handout 3: Euler's Formula on the Torus

This handout mirrors the combinatorial methods used in the Buckyball activity to prove that all Buckyballs have exactly 12 pentagon faces. Similar results can be found for "Bucky tori," where we use only pentagon, hexagon, and some higher n-gon faces. For example, if using only pentagons, hexagons, and heptagons, students can prove that there must be the same number of pentagons as heptagons. (Remember that we're also insisting that all vertices have degree 3.) If we use octagons instead of heptagons, then there must be twice as many pentagons as octagons. (This can be verified for the belcastro example given in Handout 2.)

Question 1. This problem is a lot easier to do if the students did the first page of Handout 2, where they make a fundamental domain drawing of this torus. They should get $V = 16$, $E = 32$, and $F = 16$, giving $V - E + F = 0$.

Question 2. This double-holed torus may seem a bit strange, since there are regions on the surface which look flat but which also have an edge going across them. If such edges are removed, then it will reduce the count for E and for F by one each, so it won't change the $V - E + F$ count.

In any case, students should get $V = 28$, $E = 60$, and $F = 30$, giving $V - E + F = -2$.

Thoughtful students might want to draw this double-holed torus on a fundamental domain, which would have to be drawn on an octagon (with appropriate sides identified). That isn't touched upon in this handout but would make an excellent side project or homework problem, especially for a topology course in which classification of surfaces is a goal.

Question 3. At this point the students have three data points:

surface	$V - E + F$	genus, g
sphere	2	0
torus	0	1
2-holed torus	−2	2

From this students should be able to conjecture that $V - E + F = 2 - 2g$, whereupon a formal proof can then be investigated in a topology (or graph theory) course.

Students may, however, have a difficult time with Question 3 if they don't clearly see what they should be striving for: a formula of the form $V - E + F =$ BLAH, where BLAH should be some expression with g in it. Student groups that seem to be floundering on this problem should be told that this is the goal.

Question 4. Again, students who did the Buckyball activity proving that $F_5 = 12$ should have no trouble with this activity. Since $3V = 2E$, we can rewrite Euler's formula for the torus as

$$F - \frac{1}{3}E = 0.$$

Then we use the facts $F_5 + F_6 + F_7 = F$ and $5F_5 + 6F_6 + 7F_7 = 2E$ to convert this equation to

$$F_5 + F_6 + F_7 - \frac{1}{3}\left(\frac{5F_5 + 6F_6 + 7F_7}{2}\right) = 0$$

$$\Rightarrow 6F_5 + 6F_6 + 6F_7 - 5F_5 - 6F_6 - 7F_7 = 0$$

$$\Rightarrow F_5 - F_7 = 0.$$

Thus, we must have the same number of pentagons as heptagons.

Question 5. The same exact method using octagons instead of heptagons gives $F_5 - 2F_8 = 0$, giving us twice as many pentagons as octagons.

Question 6. Part of the value of a "generalize these results" problem is to encourage the development in students of enough mathematical maturity to know, first of all, what such a generalization would mean and secondly how to go about it. This is why Question 6 is deliberately vague. The only help students might need is with setting up the problem, but instructors should resist doing this for students. Making the jump from abstraction to specific model is the real pedagogical goal of this question.

If we use only pentagons, hexagons, and n-gons to make a PHiZZ torus and do the same combinatorics as above, then we get

$$F_5 - (n-6)F_n = 0.$$

So, for every n-gon we'll need $n - 6$ pentagons.

Other projects

Once people get familiar with making PHiZZ tori, a wide range of possibilities are opened. Making "Bucky tubes" using only pentagons and hexagons is easy, and studying tori gives one the tools to make bends in these tubes. With these tools in hand, all sorts of Bucky plumbing can be performed to make spirals, n-holed tori of various shapes, and even strange things like Klein bottles. (The non-orientability of this makes it a particular challenge, but it can be done!) Searching the web for "PHiZZ unit" will reveal pictures of many such projects.

The only drawback is that such projects typically require hundreds of PHiZZ units. Still, every once in a while you will encounter a math major who gets completely obsessed with making large PHiZZ unit structures.

Do feel free to email me pictures of any interesting PHiZZ unit objects that either you or your students develop!

Activity 19
MODULAR MENGER SPONGE

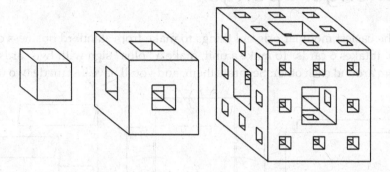

For courses: fractal geometry, discrete math, combinatorics, math for liberal arts

Summary

Students are taught the business card cube modular and paneling, which is probably one of the easiest modular designs on the planet. Students are then asked to, in groups, make a Level 1 iteration of the Menger Sponge. The handout asks them to calculate the number of cards needed to make a Level 1, 2, 3, and *n* sponge.

Content

This is really an introduction to fractals in disguise. The calculations require solving a finite geometric series and understanding the concept of self-similarity.

Handout

Page one shows how to make the basic unit and presents the activity of making a Level 1 Menger Sponge. The second page, if desired, poses the question of calculating the number of units needed to make bigger Sponges and is suited for an upper-level discrete math or combinatorics class.

Time commitment

Teaching the unit takes almost no time, but students will need 10–15 minutes to construct their first cube. Discovering how to panel cubes and make two cubes lock together will also take 15 minutes or so. Therefore, the first page of the handout may take 40 minutes total.

The combinatorial questions on the second page are meant for a combinatorics class and may take some time. This could be started with 20 minutes of class time and then finished for homework.

Business Card Cubes and the Menger Sponge

One of the easiest modular origami things to make from standard business cards is a cube. It takes 6 cards. To make a unit, make a "plus" sign with two cards and bend them around each other. Separate them, and you'll have just made two units!

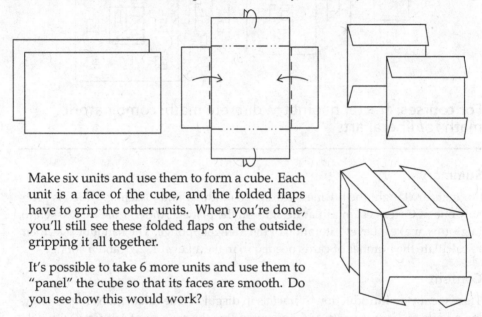

Make six units and use them to form a cube. Each unit is a face of the cube, and the folded flaps have to grip the other units. When you're done, you'll still see these folded flaps on the outside, gripping it all together.

It's possible to take 6 more units and use them to "panel" the cube so that its faces are smooth. Do you see how this would work?

Two (unpaneled) cubes can be locked together along a face by making the folded flaps grip into each other. This allows you to build structures with these cubes.

Activity: Working in groups, make a "Level 1" **Menger Sponge**. A Menger Sponge is a fractal object made by starting with a cube (Level 0), then taking 20 cubes and making a cube frame with them (Level 1), and then taking 20 of these frames and making a bigger cube frame with them (Level 2), and so on. If we scale the model down after each iteration (so it remains the same size throughout), in the infinite case we'll get what is known as Menger's Sponge.

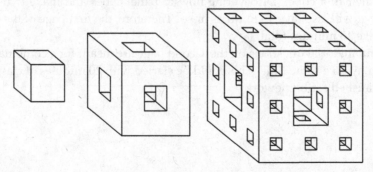

How many business cards will it take to make a Level 1 Sponge? With paneling?

Question 1: Let U_n = the number of business cards needed to make an unpaneled Level n Menger's Sponge. So $U_0 = 6$.

  Compute values for U_1, U_2, and U_3. Find a closed formula for U_n in terms of n.

Question 2: Let P_n = the number of business cards needed to make a **paneled** Level n Menger's Sponge. So $P_0 = 12$.

  Find P_1, P_2, and P_3. Can you find a formula (not necessarily closed) for P_n in general? How about a closed formula?

SOLUTION AND PEDAGOGY

A large stash of business cards will be needed for this activity, as each student will want dozens of cards. Students will certainly want to make a paneled cube of their own, which takes 12 cards. Making a Level 1 Sponge, without paneling, takes 120 units, so students really should work in groups, and a large supply of cards will be needed. (See the "Where to Find Paper" section of the introductory guide of this book for tips on where to get lots of business cards.) However, folding this modular unit is amazingly easy, and several dozen units can be folded very quickly. So it is reasonable that student groups will be able to make a Level 1 Sponge in one class period.

The instructions leave it up to the students to figure out

(1) how to make a cube from the units,

(2) how to "panel" them, and

(3) how to make two unpaneled cubes lock together.

For (1), make sure that students are leaving the "flaps" on the *outside* of the cube. If they are tucked inside, it won't stay together. Other than that, the hardest part is holding the units together as the last one is inserted. Making the folds *sharp* helps. Again, students may find it easier to do this in groups of two (more pairs of hands!) until they get the hang of it. The picture on the handout should be a big help.

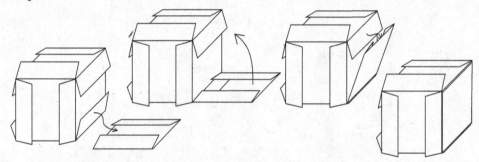

For (2), conceptually the process of paneling is pretty easy—just let the flaps of a new unit grip the flaps of a side of the cube in a perpendicular fashion (see above), but actually doing this can be tricky. It turns out to be easier to hook in one side of the panel, and then open up (slightly) one side of the cube to lock the other side of the panel. Paneling a cube has the advantage of making it *very* stable and strong.

The idea behind (3) is exactly the same as paneling, but the result is two cubes locked onto one another along a face. This is also very stabilizing, making any structure made of such cubes mighty solid.

It is also challenging to put paneling in all the interior faces of the Level 1 Sponge. Students will discover that if they want to panel it (which can be very attractive, especially if colored business cards can be found), they'll need to panel the inside faces before the outer cubes are locked in place.

Be sure to let students discover the process of making a Level 1 Sponge themselves. It can either be easy and straightforward (if they plan ahead and build it from the "inside-out") or very frustrating (if, for example, they build the outside cubes first and then try to panel the inside parts last). Planning how to construct the object helps students understand the structure of the Sponge and will provide insight on the computational questions on the flipside of the handout. A math for liberal arts or other low-level class will likely not consider the second page of the handout, as the combinatorial questions considered there are fairly challenging. But, it should be right at the level of students in a combinatorics or discrete math class for math or computer science majors.

Instructors in any class using this activity should be forewarned that it is normal for students to become addicted to making business card cube structures. During the beta-testing phase in the creation of this book, I received reports from faculty at Albion College, Davidson College, and Loyola Marymount University about how Level 2 or even Level 3 Sponges were being attempted by students, collaboratively building them in common spaces or department lounges. Incidentally, the first person to make a Level 3 Sponge out of business cards (and perhaps the only person to get all the paneling perfect) was Jeannine Mosely. Her Business Card Menger Sponge Project (see [Mos]) took many years to complete, weighs over 150 lbs, and required structural engineering problems to be overcome before success was achieved. As Dr. Mosely states, a Level 4 sponge would require over a million cards, would weigh over a ton, and thus wouldn't be able to support its own weight. Do not attempt to make a Level 4 Sponge.

Question 1

$U_0 = 6$, and the Level 1 Sponge is literally made of 20 cubes. So $U_1 = 6 \times 20 = 120$. The Level 2 Sponge will be made of 20 Level 1 Sponges, so $U_2 = 120 \times 20 = 2,400$. $U_3 = 48,000$. In general, the closed forumula is $U_n = 6 \times 20^n$.

Question 2

$P_0 = 12$, and P_1 is not nearly as easy to compute as its U_n counterpart. There are several ways to think about this, but it's more valuable to approach the problem in a way that will generalize. For example, here's one way that doesn't generalize:

$$P_1 = U_1 + \text{(panels for the 8 corner cubes)} + \text{(panels for the 12 edge cubes)}$$
$$= 120 + 8 \times 3 + 12 \times 4$$
$$= 120 + 24 + 48 = 192.$$

But then computing P_2 doesn't follow from this approach, since there are more than just corner and edge cubes in the Level 2 Sponge.

A more elegant approach is to think of P_n as 20 copies of paneled, Level $n-1$ cubes, but wherever two Level $n-1$ cubes are locked together, those sides won't need paneling. So, we just need to keep track of the places where we *won't* need paneling and subtract that number of panels. Here's how we could have done that to compute P_1:

$$P_1 = (8 \text{ corner } P_0 \text{ cubes}) + (12 \text{ edge } P_0 \text{ cubes})$$
$$= 8(P_0 - 3 \text{ panels not needed}) + 12(P_0 - 2 \text{ panels not needed})$$
$$= 8(P_0 - 3) + 12(P_0 - 2) = 8 \times 9 + 12 \times 10 = 192 \text{ units.}$$

Similarly we get

$$P_2 = (8 \text{ corner } P_1 \text{ cubes}) + (12 \text{ edge } P_1 \text{ cubes})$$
$$= 8(P_1 - 3 \times 8 \text{ panels not needed}) + 12(P_1 - 2 \times 8 \text{ panels not needed})$$
$$= 8(P_1 - 24) + 12(P_1 - 16) = 8 \times 168 + 12 \times 176 = 3456 \text{ units.}$$

Also,

$$P_3 = (8 \text{ corner } P_2 \text{ cubes}) + (12 \text{ edge } P_2 \text{ cubes})$$
$$= 8(P_2 - 3 \times 8^2) + 12(P_2 - 2 \times 8^2)$$
$$= 66{,}048 \text{ units.}$$

This suggests a general recursive formula:

$$P_n = 8(P_{n-1} - 3 \times 8^{n-1}) + 12(P_{n-1} - 2 \times 8^{n-1}) = 20P_{n-1} - 6 \times 8^n.$$

In fact, now that you see this recurrence, you might be able to see a more simple justification of it (if you didn't see it already!): To get P_n we need to take 20 Level $n-1$ paneled cubes (which each take P_{n-1} cards), and then we need to subtract the paneling that we don't need. Each of the 12 edge-positioned Level $n-1$ cubes has two sides that won't require paneling (so $12 \times 2 = 24$), and then each of these sides will be facing the side of a corner cube that won't need paneling either. So that's 48 sides total that won't need paneling. Now, the side of a Level $n-1$ cube will need 8^{n-1} cards to panel it, so we need to subtract $48 \times 8^{n-1} = 6 \times 8^n$, giving the desired recurrence.

   This recurrence can be solved (to get a closed formula) using generating functions: Multiply the equation by x^n and sum over all $n \geq 1$ to get

$$\sum_{n=1}^{\infty} P_n x^n = 20 \sum_{n=1}^{\infty} P_{n-1} x^n - 6 \sum_{n=1}^{\infty} 8^n x^n.$$

Our generating function will be $G(x) = \sum_{n=0}^{\infty} P_n x^n$. Plugging this in and using $\sum_{n=0}^{\infty} (8x)^n = 1/(1-8x)$ gives

$$G(x) - P_0 = 20xG(x) - 6\left(\frac{1}{1-8x} - 1\right)$$

$$\Rightarrow G(x)(1-20x) = 12 - \frac{6}{1-8x} + 6 \Rightarrow G(x) = \frac{18}{1-20x} - \frac{6}{(1-8x)(1-20x)}.$$

Partial fractions are needed to break up the last term, so we set

$$\frac{6}{(1-8x)(1-20x)} = \frac{A}{1-8x} + \frac{B}{1-20x},$$

which gives $6 = A(1-20x) + B(1-8x)$. Using a standard Calc II trick, we can let $x = 1/8$ to give us $A = -4$ and $x = 1/20$ to give $B = 10$. Thus, we have our generating function:

$$G(x) = \frac{8}{1-20x} + \frac{4}{1-8x} = 8\sum_{n=0}^{\infty} 20^n x^n + 4\sum_{n=0}^{\infty} 8^n x^n$$

and so $P_n = 8 \times 20^n + 4 \times 8^n$.

Undoubtedly there are other ways to compute this, perhaps more easily than the above method. However, since recurrence relations and generating functions are standard material for an undergraduate combinatorics course, this activity can provide a surprising and accessible application of these methods.

Follow-up/senior project

Students studying fractal geometry who have taken a combinatorics course would be prepared to investigate the problem of computing the surface area and volume of the Menger Sponge. This object, like many fractals, exhibits counterintuitive behavior in this regard: The "infinite iteration" of Menger's Sponge has zero volume but infinite surface area.

Remember that when performing such an analysis, each iteration of the Sponge needs to be at the same scale. That is, if we assume that the Level 0 Sponge (cube) has side length 1, then so should all Level n Sponges. (So a Level 1 Sponge will have volume $20 \times (1/27)$.) Continuing this, we see that the volume of a Level n Sponge is $(20/27)^n$, which goes to zero as n goes to infinity.

The number of panel units, $P_n - U_n$, can be used to compute the surface area of a Level n Sponge, and taking the limit of this shows that the surface area goes to infinity.

Activity 20
FOLDING AND COLORING A CRANE

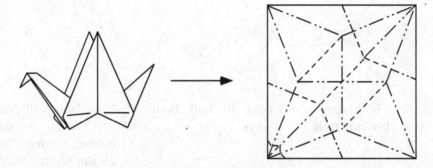

For courses: discrete math, graph theory, math for liberal arts

---

Summary

Students are taught the flapping bird (a more simple version of the traditional Japanese crane) model. They are then asked to unfold their model and draw the crease pattern. Then they are asked to color the regions of the crease pattern so that no two neighboring regions receive the same color, and they should use as few colors as possible. What do they think will happen when we refold the model? What does this tell us?

Content

While this activity touches upon the beginnings of the field of "computational origami," it is also a simple graph coloring exercise. Giving a purely theoretical proof is a good basic graph theory exercise.

Handout

Only one, containing instructions for folding the flapping bird and the activities for drawing the crease pattern and coloring it.

Time commitment

Teaching the crane (flapping bird) should take 15–20 minutes, and drawing the crease pattern can take some time. After that, coloring and studying the coloring does not take long; 45 minutes for the whole activity is a good bet.

Folding a Flapping Bird (Crane)

Begin with a square piece of paper.

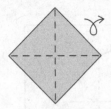 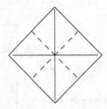 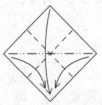

(1) Crease both diagonals. Then **turn over**.

(2) Fold in half both ways.

(3) Now bring all corners down to the bottom, using the creases just made,...

(4) ...like this. This is called the **preliminary base**. Bisect the two angles at the open end.

(5) Then fold the top point down.

(6) Undo the last two steps.

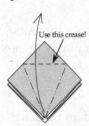  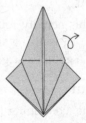

(7) Now do a **petal fold**: lift one layer of paper up, using the indicated crease as a hinge,...

(8) ...like this. Bring the point all the way up. The sides will come to the center. Flatten...

(9) ...like this. Turn over.

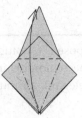
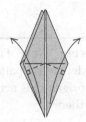

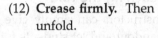

(10) Now do the same petal fold on this side.

(11) This is called the **bird base**. Fold the bottom two flaps up. (These will become the head and tail.)

(12) **Crease firmly.** Then unfold.

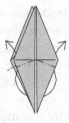

(13) Now refold the last creases, but this time make them **reverse fold** through the layers. (See the next...

(14) ...picture.) Lastly, reverse fold the head.

(15) You're done with the flapping bird!

This is an example of a **flat origami model**, since the finished result can be pressed in a book without crumpling.

Activity 1: Carefully **unfold** your bird and draw with a pen the crease pattern for this model. Make sure to draw **only** those creases that are actually used in the finished model, not auxiliary creases made along the way.

Activity 2: Then take your crease pattern and **color the faces** with as few colors as possible. That is, color the regions in between crease lines following the rule that no two regions that border the same crease line can get the same color (just like when coloring countries on a map). What's the fewest number of colors that you can use?

Activity 3: What will the coloring look like when you refold the model? Make a conjecture before you fold it back up to see what happens. Will this happen for **every** flat origami model? Proof?

SOLUTION AND PEDAGOGY

This is a fairly simple activity with a big "Wow" factor. Its purpose is for students to discover that all flat origami model crease patterns are 2-face-colorable in the graph theory sense. The "proof by origami" is actually quite elegant, although it can also be proven purely by graph theory.

Teaching the fold

Instructors can simply give students the diagrams for the flapping bird in the handout and let students follow them at their own pace. (Working in groups to help each other out is a very good idea.) Or instructors can lead the class in folding it step by step. In either case instructors should fold this model themselves several times to become very comfortable with the more tricky petal fold (steps (7)–(8)) and reverse fold (steps (13) and (14)), as these always give some students trouble.

The instructions depict paper that is colored on one side and white on the other. Traditional origami paper (kami) has this property, but it is neither needed nor desired for this activity. Plain white squares of paper are better so that students can easily draw the crease pattern and color the regions. Cutting white photocopy paper into squares makes a good size for this model and activity.

Although the handout doesn't mention it, there is a reason why this model is called the Flapping Bird. If you pinch the base of the neck with one hand and gently pull on the tail with the other hand, you can make the wings flap. A newly-folded model needs to be "coaxed" into allowing the flapping mechanism to work, and then it should flap easily. This nice side effect has no bearing on the mathematics of this activity whatsoever.

The activities

Students may be reluctant to unfold their creation, but if you tell them to just follow the origami instructions in reverse, it'll be easier for them. Drawing the crease pattern is rather tricky, only because it's easy to draw a crease line that is not used in the final model. Emphasize that only crease lines that are used in the end may be drawn.

The result should look as follows (I've indicated which creases are mountains and which are valleys, but the students need not keep track of that):

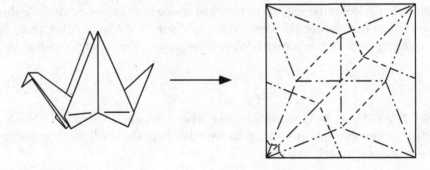

If it helps, students can try drawing the creases used as they unfold the model, step-by-step, or students can draw on the creases with a magic marker or felt tip pen while the paper is still folded. I've also found it helpful to show them what it should look like once everyone has had a chance to begin or almost finish their crease pattern. That way they can make sure it's correct for the coloring part of the activity.

Only two colors are needed to color the regions of the crease pattern. Since each crease line borders two regions of different colors, when the crease is refolded it will make the two colors face in opposite directions. Thus a 2-colored flat origami crease pattern will, when folded, result in a model that is all one color on one side and all the other color on the other side.

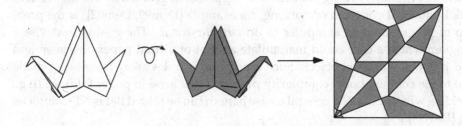

In fact, this offers a slick proof that all flat origami model crease patterns are 2-face-colorable: Fold it up to get a flat object. Since it's flat, each region of the paper will face in one of two different directions, say left and right. Color every region that faces left white and every region that faces right grey. When the model is unfolded, the crease pattern faces will be colored in only two colors and no two neighboring faces will have the same color.

One can also prove this using only graph theory. First argue that all vertices in the interior of the paper of a flat model have even degree. (This is not so easy for undergraduates to prove rigorously, so you may want to allow some leeway here.) Then, if we consider the crease pattern to be a planar graph, where the boundary of the square also contributes edges to the graph, the only odd degree vertices would possibly be on the paper's boundary. Create a new vertex, v in the "outside face" and draw edges from it to all the odd degree vertices on the paper's boundary. Graphs always have an even number of vertices of odd degree, so the degree of v is even, and the new graph that we've created has all vertices of even degree. Planar graphs with all vertices of even degree have duals that are bipartite. So the dual is 2-vertex-colorable, meaning that the crease pattern with the vertex v is 2-face-colorable. Removing the vertex v then gives a 2-face-coloring of the original crease pattern.

In a graph theory course, this proof can be used to reinforce some basic concepts (duality, bipartite graphs, degrees of vertices). Developing such a "pure graph theory" proof can be a great exercise for students.

One way to speed things up is to adopt a revised order for this activity: (1) Fold the crane. (2) With the paper still folded, have students color the regions of paper between the creases so that all faces facing one direction are grey, say, and all regions facing the other direction are white. (3) Then have them unfold the model and explain why this results in a proper face coloring of the crease pattern. Coloring the regions while the paper is folded can be tricky, especially for the small regions around the head. But this alternate approach can cut down the time used by a lot.

Why do we care?

As mathematicians, this 2-colorability result has inherent appeal. Many math students will appreciate this, but there's another motivation at play here. One of the big open areas in the new field of "computational origami" (which, yes, really is a new and rapidly growing subfield of computational geometry—see the work of Erik Demaine if you need convincing, for example [Dem99, Dem02]) is the problem of programming a computer to do *virtual origami*. The goal is to devise a program where a user could manipulate a sheet of virtual paper on-screen and fold it into any origami object. Such a program is miles away from being made due to the computational complexity problems that arise in paper folding. (E.g., deciding whether or not a general crease pattern can be folded flat is NP-complete; see [Bern96].)

This 2-colorability result gives us a very fast way for a computer to be able to determine the direction in which each region of a flat origami crease pattern will face when folded. Thus, this result is very helpful to those studying the computational side of origami.

Activity 21
EXPLORING FLAT VERTEX FOLDS

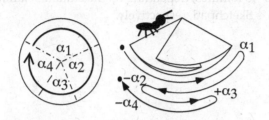

For courses: geometry, discrete math, combinatorics, math for liberal arts, intro to proof, modeling

Summary

Students are asked to fold, from numerous small pieces of paper, lots of flat vertex folds—origami models that fold flat and whose crease pattern contains only a single vertex, say, in the center of the paper. The mission: find patterns, make conjectures, and find proofs and counterexamples.

Content

The conjectures that the students make, and their proofs, will involve some basic geometry, combinatorics, and careful reasoning. Thus this could be used in a geometry or combinatorics course as an early activity to emphasize the process of exploration, conjecture, and proof. The overhead for this is minimal, so it could also be used in a math for liberal arts or intro to proofs course. Further, this is a fine example of taking a physical situation, studying it, and creating the language, notation, and theory that you need to model it mathematically. Thus, this would fit right into a mathematical modeling class. The things conjectured here also form the basics of flat origami theory.

Handouts

The first handout is deliberately simple and open-ended. The main idea is to get students to make their own conjectures and look for either counterexamples or proofs. There are many conjectures that can be made about flat vertex folds (described in the solutions section), so the handout doesn't provide any hints. It's best to let the students discover what they will here.

The second and third handouts are activities for modeling flat vertex folds using either Geogebra or Geometer's Sketchpad. The specific purpose is to give students an experimental way to discover Kawasaki's Theorem.

Time commitment

The first handout is very open-ended and can take a whole class period or several class periods, depending on how instructors want to do it. The second or third handout can take 30–40 minutes, depending on the students' familiarity with Geogebra or Geometer's Sketchpad, respectively.

Exploring Flat Vertex Folds

Activity: Take a square piece of paper and make, at random, a single vertex crease pattern that folds flat. Place the vertex near the center of the paper (not on the paper's boundary—that doesn't count), make some crease lines coming out of it, and then add more to make the whole thing fold flat. Some examples are shown below. Make lots of your own.

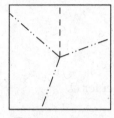 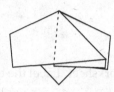 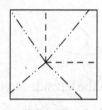 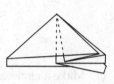

The question is, "What's going on here?" Are there any rules that such flat vertex folds follow? **Your task** is to formulate as many conjectures as you can about how such folds work.

If you come up with a conjecture, write it on the board to see if others in the class agree or if anyone can find a counterexample. Or, better yet, see if anyone can actually give proofs of your conjectures!

Flat Vertex Folds in Geogebra

To simulate a flat vertex fold using Geogebra, do the following:

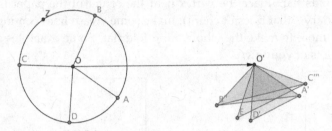

(1) Make a circle on the left side of your worksheet. Label the center O.

(2) Make four points on the circle, A, B, C, and D.

(3) Construct segments between these points and O. This circle and these lines will be your unfolded piece of paper.

(4) On the right side of your worksheet make a point O'. Use the **Vector between Two Points** tool to make a vector from O to O'.

(5) Now use the **Translate Object by Vector** tool to translate A to a new point A' in the direction of the vector from step (4). Do the same with points B, C, and D to make points B', C', and D'.

(6) Use the **Polygon** tool to make $\triangle O'A'D'$. This triangle will be the start of our folded paper.

(7) We now will reflect (fold) the points B', C', and D' about the crease line $O'A'$. Use the **Reflect Object about Line** tool to reflect each point, one at a time, to get new points B'', C'', and D''.

(8) Now use the **Polygon** tool to make $\triangle O'A'B''$.

(9) Now reflect (fold!) the points C'' and D'' about the line OB'' to do the third fold. This will make new points C''' and D'''.

(10) Use the **Polygon** tool to make $\triangle O'B''C'''$.

(11) Now reflect D''' about the crease line $O'C'''$ to make a new point E. (Geogebra will use a new letter because it doesn't like D''''.)

(12) Use the **Polygon** tool to make the last triangle of the folded paper, $\triangle O'C'''E$.

(13) Now use the **Show / Hide Object** tool to hide the points B', C', C'', D'', and D''' because we no longer need them.

Exercise: Does the last point you made, E line up with point D'? If so, then the crease lines you made on the left can fold flat. If they do not, then move the points on the left circle until they do. Use Geogebra to measure the angles $\angle AOB$, $\angle BOC$, $\angle COD$, and $\angle DOA$. What can you conjecture about these angles when the creases fold flat?

Flat Vertex Folds on Geometer's Sketchpad

To simulate a flat vertex fold on Geometer's Sketchpad, do the following:

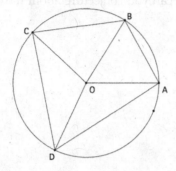 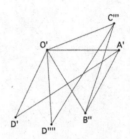

(1) Make a circle on the left side of your worksheet. Label the center O.

(2) Make four points on the circle, A, B, C, and D.

(3) Construct segments between these points and O. Also construct segments between A, B, C, and D in order to make a quadrilateral (as shown above).

(4) Select the quadrilateral, points A, B, C, D, and O, and the segments at O, and select **Translate** from the **Transform** menu. Choose Rectangular coordinates and make the horizontal and vertical distance be 12 cm and 0 cm, respectively.

(5) You now have a second copy of the quadrilateral "paper" with creases. Select the text tool and click on all the points of this copy to see what they are (A', B', C', D', O').

(6) We now will reflect parts of this copy about the creases to make it fold up. Select segment $O'A'$ and choose **Mark Mirror** from the **Transform** menu.

(7) Now select segments $A'B'$, $B'C'$, $C'D'$, $O'B'$, $O'C'$, and $O'D'$ and points B', C', and D'. With all this selected, choose **Reflect** from the **Transform** menu.

(8) You've just make $\triangle O'A'D'$ fixed and reflected the rest of the paper about crease $O'A'$! Now we want to hide the parts that we had previously selected. Under the **Edit** menu choose **Select Parents** and then *unselect* segments $O'A'$ and $O'D'$ and point D'. Then, under the **Display** menu choose **Hide Objects**.

(9) Use the text tool to click on the new points to see what they are (B'', C'', D'').

(10) Now select segment $O'B''$ and do **Mark Mirror**.

(11) Select segments $B''C''$, $C''D''$, $O'C''$, and $O'D''$ and points C'' and D''. Then do **Reflect**.

(12) Again, do **Select Parents**, *unselect* segment $O'B''$, and then **Hide Objects**.

(13) Label the points again, select segment $O'C'''$, and do **Mark Mirror**.

(14) Select $C'''D'''$ and $O'D'''$ and **Reflect**. Then **Hide** $C'''D'''$, $O'D'''$, and D'''.

Exercise: Does the last point you made, D'''' line up with point D'? If so, then the crease lines you made on the left can fold flat. If they do not, then move the points on the left circle until they do. Use Geometer's Sketchpad to measure the angles $\angle AOB$, $\angle BOC$, $\angle COD$, and $\angle DOA$. What can you conjecture about these angles when the creases fold flat?

SOLUTION AND PEDAGOGY

Instructors will have to practice making lots of flat vertex folds themselves before leading this activity. The reason for this is that most people enter this kind of free-form exploration of paper folding with preconceived notions as to what paper can do. Instructors may have such notions as well.

These notions can include things such as, "crease lines must go all the way through the paper" (they do not—see the right figure below for an example different from the ones on the handout) or "one can't have too many mountain creases (or valley creases) in a row." (The left figure below shows how lots of mountains can be consecutive.)

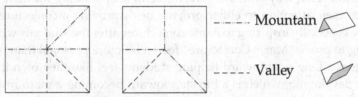

Folding lots of examples *in groups* so that students can share their folds with each other should eliminate such preconceived notions. (Note that these folded vertices must be in the *interior* of the paper.)

However, another problem that many students have is in making "whimpy creases." In these flat vertex folds it is fairly important that creases be made sharp. Soft, ambiguous creases can be so inaccurate that students will often think that they have a counterexample to someone's conjecture when in fact they do not.

Conjectures

So what kinds of conjectures might students develop? Here's a list:

(1) Flat vertex folds always have even degree (number of creases).

(2) The angle between two consecutive creases in a flat vertex fold is always $\leq 180°$.

(3) If we stab the folded vertex somewhere reasonable (i.e., not near the boundary of the paper and not directly on a crease), then we'll always get an even number of layers of paper at that point.

(4) The number of mountain creases and the number of valley creases always differ by 2 in a flat vertex fold.

(5) If α_1, α_2, and α_3 are consecutive angles in a flat vertex fold and if $\alpha_1 > \alpha_2$ and $\alpha_3 > \alpha_2$, then the two creases separating these angles must have different mountain-valley parity.

(6) If $\alpha_1, \alpha_2, \ldots, \alpha_{2n}$ are the angles, in order, between creases in a flat vertex fold, then $\alpha_1 - \alpha_2 + \alpha_3 - \cdots - \alpha_{2n} = 0$.

(7) (Using the same notation as (6)) $\alpha_1 + \alpha_3 + \cdots + \alpha_{2n-1} = \alpha_2 + \alpha_4 + \cdots + \alpha_{2n} = 180°$.

(8) (Harder) If we draw crease lines meeting at a vertex with consecutive angles satisfying $\alpha_1 - \alpha_2 + \alpha_3 - \cdots - \alpha_{2n} = 0$, then the vertex will fold flat.

Of course, students may conjecture other things, like "You can't have all mountains or all valleys," which while true, are pretty simple. Also, all of the above conjectures are true; students may develop some false ones. Such conjectures, or ones that you never thought of before, should be treated with equal seriousness.

I like to keep a running list of the conjectures on the board as students explore flat vertex folds. That way students can choose to keep looking for more conjectures or turn their attention to either proving or disproving a conjecture on the list. It can be especially inspiring to name conjectures after the students who make them. Trying to prove "Max's Conjecture" feels a lot more personal than "Conjecture 2." It goes a long way toward helping students feel like they own the math that they're developing, which is a big step toward becoming a math researcher. Also, you can probably see how this activity could take several class periods if you like. A running list of conjectures can be assigned for homework, in a Moore method-like approach to it all.

As mentioned earlier, it's best for the students to come up with these conjectures themselves. In the theory of flat origami, conjectures (4) and (6)–(8) above are probably the most significant. They are known as Maekawa's and Kawasaki's Theorems, respectively [Kas87], though they were also discovered by Justin [Jus84, Jus86], and (6) was discovered independently by other people as well. (See [Rob77] and [Law89]. But if one of them is missed by the class there's no real harm done. (Unless you plan on also doing subsequent activities on flat origami or the matrix models, in which case you may need the class to know Kawasaki's Theorem.)

In fact, it is very likely that students will not see the angle condition needed for (6)–(8) above. Since this is an important result for some of the other activities, I included a handout that shows students how they might simulate a four-valent flat vertex fold on Geogebra or Geometer's Sketchpad. The idea is this: Flat folds require that each crease line acts like a reflection of the plane. So, we create a degree-4 vertex in Gegebra and imagine that it has been cut along one of the crease lines (segment OD on the handouts). Then, we use the reflection tools of the geometry software to show what folding along the other crease lines would look like. If the two cut ends of the paper line up, then the four creases make an foldable flat vertex crease pattern. If they do not line up, then the creases do not fold flat.

The purpose is to allow students to measure the angles between the creases so that they can generate data of which angles will work for a flat vertex fold. This gives the students a chance to actually conjecture Kawasaki's Theorem in the degree-4 case, which can then lead to the general theorem.

Still, it is important for the students to see that there is real mathematics going on with these little folded pieces of paper, so some tactful hints can be suggested.

For example, oftentimes students don't even think about considering possible patterns among the mountain and valley creases. The handout actually shows some mountains and valleys on the sample vertex folds, so that is a subtle hint. If the students don't pick up on that, you may want to suggest out-loud to the whole class as you wander amongst the groups, "It's funny that no one is thinking about the mountain and valley creases." Then they'll start conjecturing about them.

Proofs of the conjectures

There are many ways to prove these conjectures. If this were an origami-math textbook or monograph, I would choose an elegant order in which certain results flow from one to another. But your students won't be doing it that way, so while yes, some results are more easily proven from others, it helps to know how to prove them separately as well. (Of course, this is the difference between doing research yourself and reading about it, sans scribbles and scratch work, in a publication.)

So, in no particular order, I'll list several proofs. You and your students may find more. For references in the literature, see [Hull94], [Hull02-1], and [Hull03].

Maekawa's Theorem: *Let M and V denote the number of mountain and valley creases, respectively, that meet at a flat vertex fold. Then $M - V = \pm 2$.*

Proof 1: Fold the vertex flat and imagine cutting the vertex off with scissors, leaving a flat polygonal cross-section. (See the illustration below.) Imagine a monorail traveling along this cross-section in a counterclockwise manner. Then, assuming that we're looking at the cross-section from above, every time the monorail gets to a mountain crease it will rotate 180°, and every time it gets to a valley crease it'll rotate −180°. When it gets back to where it started it will have rotated a full 360°. So,

$$180M - 180V = 360 \Rightarrow M - V = 2.$$

If we had looked at the vertex "from below," we would have gotten −2. □

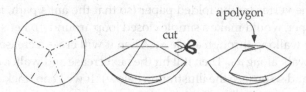

a polygon

cut

Proof 2: (Jan Siwanowicz, HCSSiM class of 1993, developed this proof.) If n is the number of creases, then $n = M + V$. Fold the paper flat and consider the cross-section obtained by cutting off the vertex; the cross-section forms a flat polygon. If we view each interior 0° angle in this polygon as a valley crease and each interior 360° angle as a mountain crease, then the sum of the polygon's interior angles gives $0V + 360M = (n - 2)180 = (M + V - 2)180$, which gives $M - V = -2$. If we reverse the roles of mountain and valley creases (this corresponds to flipping the paper over), then we get $M - V = 2$. □

Even Degree Theorem: *Every flat vertex fold has even degree.*

Proof using Maekawa: Let the number of creases at the vertex be $n = M + V = 2V + M - V = 2V \pm 2 = 2(V \pm 1)$, which is even. □

Proof using coloring: If the students have done the Folding and Coloring a Crane activity, they'll know that our flat vertex folds are all 2-face-colorable. This immediately gives us that there are an even number of creases. □

Stand-alone proof: Using the monorail approach as seen in Proof 1 of Maekawa, keep track of the times the monorail is traveling left or right by a sequence of Ls and Rs. Since each crease is folded flat, this sequence will alternate L and R. If we stop keeping track of these once we arrive at the region where the monorail started, we get the same number of Ls as Rs, so the sequence has even length. The length of this sequence equals the degree of the vertex. □

Big-Little-Big Angle Theorem: *Suppose that in a flat vertex fold we have a sequence of consecutive angles α_{i-1}, α_i, and α_{i+1} with $\alpha_{i-1} > \alpha_i$ and $\alpha_i < \alpha_{i+1}$. Then the two crease lines in between these three angles cannot have the same mountain-valley parity.*

Proof: For the sake of contradiction, suppose that the two creases are both valleys or both mountains. Then, when they were folded, we would have both big angles α_{i-1} and α_{i+1} covering up smaller α_i on the same side of the paper. This is impossible to do without the paper intersecting itself. Thus the two crease lines must have different mountain-valley parity. □

Kawasaki's Theorem: *A vertex fold v folds flat if and only if the alternating sum of the consecutive angles between the creases at v equals zero.*

Proof: Let v be a flat vertex fold with consecutive angles between the creases $\alpha_1, \ldots, \alpha_{2n}$. Fold the vertex flat and imagine an ant being dropped on a crease, who then walks around the vertex on the folded paper (so that the ant's path, if marked on the unfolded paper, would make a simple closed loop around v). Let's assume that the ant starts by walking through angle α_1. Then it will cross a crease line, switch directions, and walk along α_2. Then it'll hit the next crease and walk α_3 in the same direction as α_1, and so on (see the illustration below). If we keep track of the angles that the ant swings out, we'll get an alternating sum $\alpha_1 - \alpha_2 + \alpha_3 - \cdots - \alpha_{2n}$. At the end the ant should come back to where it started, so this sum should equal 0.

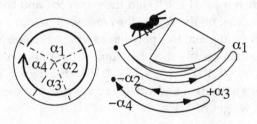

For the converse, we assume that $\alpha_1 - \alpha_2 + \alpha_3 - \cdots - \alpha_{2n} = 0$, and we want to show that the vertex can fold flat. We'll do this by generating a mountain-valley assignment for the creases at v that will not force the paper to self-intersect when folded.

Pick a crease line l of v at random and cut along this crease line, making two "loose ends" of paper where l used to be. Then, assign alternating mountains and valleys to the remaining crease lines. We can then fold these creases up, where the cross-section would look like a zig-zag pattern. Since the alternating sum of the angles is zero, we know that the two loose ends will end up aligned with each other. If we're lucky, these loose ends will have no paper in between them, whereupon we can glue them back together (which will assign either a mountain or a valley to l) and the vertex will have been folded flat. (See the illustration of this below.)

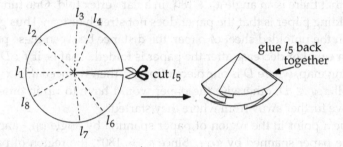

If we're unlucky, however, there will be layers in between the loose ends of l. In that case (as illustrated below) we need to look at the cross-section of our zig-zag pleats and, assuming that they go left-and-right, choose the right-most crease in this cross-section to reverse, turning it from a mountain to valley or vice versa. Doing this will place the loose ends on top of one another with no flaps of paper in between them, whereupon they can be glued back together to complete the flat vertex fold. □

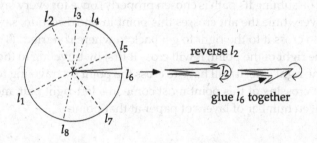

Angle Theorem: *The angle between two consecutive creases of a flat vertex fold is always $\leq 180°$.*

Proof using Kawasaki: Since the vertex folds flat, we know that $\sum (-1)^i \alpha_i = 0$. But we also know that $\sum \alpha_i = 360°$. Adding these equations together and simplifying, we get

$$\alpha_1 + \alpha_3 + \cdots + \alpha_{2n-1} = 180°, \text{ and}$$

$$\alpha_2 + \alpha_4 + \cdots + \alpha_{2n} = 180°.$$

Thus no angle in a flat vertex fold can be greater than or equal to 180°, unless it is a "trivial" vertex of degree 2, which would have angles of exactly 180°.   □

Yes, one doesn't necessarily want to consider vertices of degree 2, but some students might insist that they exist. When developing such "flat origami theory," it is sometimes convenient to allow vertices of degree 2, so don't sweat it.

Stand-alone proof: It's likely that students will try to prove this before discovering Kawasaki, so a proof without that result is needed. A proof by contradiction works quite well.

Suppose that there is an angle $\alpha_i > 180°$ in a flat vertex fold. One fundamental fact about folding paper is that the paper does not stretch or tear. Thus given any two points on the unfolded sheet of paper, the distance between these points can either remain equal or decrease after the paper is folded. That is, if $f : D \to \mathbb{R}^2$ is our flat folding map, where D is our piece of paper, then we need $d(f(x), f(y)) \leq d(x, y)$, for all $x, y \in D$, otherwise the paper would have to rip in order for the points to move further away from where they started.

So let x be a point in the region of paper spanned by angle α_{i-1} and let y be a point in the paper spanned by α_{i+1}. Since $\alpha_i > 180°$, the region of paper that contains angle α_i is not convex. Thus, if we imagine α_i's region as remaining fixed, folding the crease lines between it and angles α_{i-1} and α_{i+1} will move the points x and y further away from each other, which is a contradiction.   □

Number of Layers Theorem: *The number of layers of paper near a flat vertex fold at any point that does not intersect an edge is even.*

Proof: Using the ant-walking argument of the Kawasaki's Theorem, the ant would cross this point (assuming its path is chosen properly) once for every layer of paper at this point. Every time the ant crosses this point in one direction, say to the left, then it must also cross it to the right to get back to where it started. (That is, if the ant begins to the right of the point, it will cross it first by traveling to the left. Then, to get back to the right side, it will have to cross the point by traveling to the right.) Thus, every ant crossing of this point must come in a left-right pair, meaning that we'll have an even number of layers of paper at that point.   □

Pedagogy

The list of conjectures and proofs is only provided here because it helps a lot for instructors to know what to expect when embarking on an open-ended activity such as this. Instructors *must resist* the urge to revert into lecture mode and just present this string of conjectures and proofs to the students. This would defeat the entire purpose of the activity, which is to present students with a problem that is completely unfamiliar, easy to investigate, requires no prior knowledge, and contains deep insights to discover. In this way they can get first-hand experience with mathematical research as they look for patterns, make conjectures, and try to prove them.

There's a very good chance that students will not make all the conjectures on the list. Perhaps they'll discover a few not on the list! Teaching such an open-ended activity can be very challenging because you don't know exactly what will happen in class. It's best to not think of this as material that needs to be covered. Rather, it's the experience of wrestling with the problem and the development of conjectures and proofs that should be the main goal. If you plan on also doing more flat foldability activities, like the Impossible Crease Patterns activity or the matrix model activities in this book, then you will want to make sure that they discover Maekawa and Kawasaki, which can be done by dropping hints.

On one hand, this activity is very simple and fun, since the math involved requires no overhead. On the other hand, it is very challenging because it requires students to think like mathematicians and to "do math" in a way very different from what they probably have seen before.

Motivation may also be an issue, especially for a lower-level or "math for liberal arts" class. For such classes it would be very helpful to have them fold some actual models, like the flapping bird in the Folding and Coloring a Crane activity, before diving into individual flat vertex folds.

Bad proofs

When formulating proofs to these conjectures, there is a strong tendency for students to pursue lines of thought that can be very unfruitful. What's worse, there are certain arguments that can be made for Maekawa and Kawasaki in particular that sound very convincing to students but that are completely nonrigorous and false. This makes the proof-building part of this activity very valuable, as some conjectures are not hard to prove at all (but require sound thinking), while others can be very challenging.

Instructors should be especially on the lookout for attempts to prove Maekawa or Kawasaki by induction. Such proof attempts are in some sense doomed to fail because once you remove some crease lines from a general flat vertex fold, the result is not likely to be a flat vertex fold anymore! Still, some students will insist that, for example, everywhere there is a mountain crease there should also be a valley crease to go with it. So, for example, the most "basic" (base case) flat vertex fold is one of degree 2, which we can think of as a vertex with two mountains, say, around it. Then, any other flat vertex fold will be adding a mountain and a valley to this and then repeating, always resulting in two more mountains than valleys. Some students will swear up and down that this is a valid argument, but of course it is nonsense. (On the other hand, there are other ways in which induction might work; see the "Follow-up things" section for more info.)

The difficulty in proving these conjectures is that students have a hard time seeing where to start. There are no immediate formulas or "mathy" things to use that they can easily see. This is yet another reason why this activity can be a valuable experience for students, since often mathematicians have to face situations in which we must create the mathematical model ourselves before we can prove anything. All students have here is a folded piece of paper. Making a model by defining the angles and the mountain-valley creases is a start, but it is very difficult to generate anything that would lead to a rigorous proof without a dose of creativity, like wondering what it would be like to crawl around the vertex on the folded paper, or to cut off the vertex and look at the cross-section it reveals. Those are the keys to good proofs here. Also, the proof by monorail and ant-walking techniques can be very helpful. They offer a way to visualize what the paper is doing. Suggesting this technique (they are, after all, basically the same thing) or even offering the ant-walking proof of Kawasaki may give them ideas for proving other things.

Pedagogically, it can be very difficult for instructors to balance the need for students to develop their own proofs with the desire to move things along by giving hints. In this sense, it is almost dangerous for instructors to know the above proofs, for if you didn't know them then you would be forced to let students devise proofs on their own. After all, one of the above proofs of the Maekawa's Theorem was developed by a student (a high-school student, at that), so you never know what new approaches students might come up with.

Follow-up things

There is a wealth of directions in which students could go to pursue the subject of flat vertex folds further. This has great potential for student projects, including working on accessible open problems.

This book contains other activities (following this one) that look at some of these directions. Asking whether or not Kawasaki's Theorem can be generalized to crease patterns with more than one vertex is the subject of the Impossible Crease Patterns activity. Also, an equivalent version of Kawasaki can be made using a matrix model for flat folds, as explored in the Matrix Model of Flat Vertex Folds activity.

Students and instructors who want to know more of the full story, however, are encouraged to read my paper "The Combinatorics of Flat Folds: A Survey" [Hull02-1]. One of the things mentioned in that paper is how both Maekawa's and Kawasaki's Theorem proofs never actually use the fact that the paper we were using was flat. Thus, both of these results apply to folds on paper with different curvature. For example, if we were folding a cone where the vertex was placed at the cone point, then both of these results would still hold. In this form, one can actually prove these theorems using a careful induction argument, since removing creases can then be the same as reducing the amount of paper around the cone.

Another avenue to travel in this area is counting the number of valid mountain-valley assignments that are possible for a given crease pattern. The Folding a Square Twist activity looks at an instance of this, as does the survey paper mentioned above (see also [Hull03]).

Activity 22
IMPOSSIBLE CREASE PATTERNS

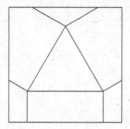

For courses: geometry, discrete math, combinatorics, math for liberal arts, intro to proof, modeling

Summary

This is really a follow-up for the previous activity on flat vertex folds, but it doesn't have to be. Students are given squares of paper with crease patterns drawn on them and asked to fold along the lines to fold the paper up into something flat. The catch is that the crease patterns are impossible to fold flat without inserting new creases. This is puzzling because each vertex will locally fold flat, but the global pattern will not. Students are asked to explain why these won't fold up.

Content

On a basic level this offers students more practice examining real-life situations and trying to analyze them mathematically. Having done the previous activity puts this one into a better context, but this activity by itself requires no overhead.

With the previous activity, however, students are poised to look more deeply. Given a single vertex crease pattern, we can easily determine whether or not it can fold flat. But a multiple-vertex crease pattern poses more difficulties, as illustrated in the impossible crease patterns of this activity. It turns out that deciding whether or not crease patterns can be folded flat in general is NP-complete. Thus, this can be an illustration to students in an analysis of algorithms course of the different contexts in which decidability and computational complexity can arise. Actually proving NP-completeness is beyond the scope of the activity, but playing with and discussing the problems with the impossible crease patterns can give an appreciation of how hard this problem can be with larger crease patterns.

Handout

The handout is minimal and is only a device by which to deliver the crease patterns. They can be cut out by either the students or the instructor ahead of time.

Time commitment

How much time this takes is entirely dependent on how many of the crease patterns you want your students to try. Each one takes only 5–10 minutes to try folding, but developing arguments as to why they don't work will probably take another 10 minutes each.

Fold Me Up

Activity: Below are some origami crease patterns. Your task is to cut them out and try to see what they can fold into. Note: You're only allowed to fold along the indicated crease lines. Adding more creases is breaking the rules. You get to decide, however, whether to make them mountains or valleys.

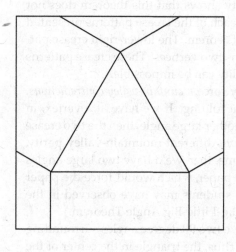

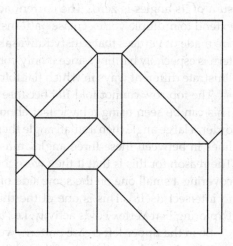

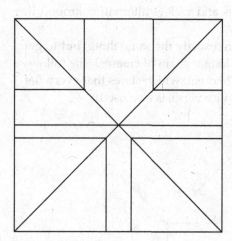

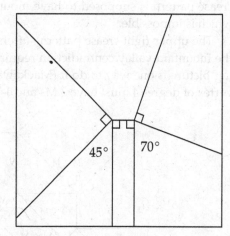

SOLUTION AND PEDAGOGY

Of course, the whole, perhaps devilish point is that all of these crease patterns are impossible to fold flat. Thus students may experience some frustration with these crease patterns until they realize that the challenge is seeing that they are impossible and then trying to figure out why.

This activity makes a great follow-up to the Exploring Flat Vertex Folds activity. In that activity, one of the main theorems that can emerge is the so-called *Kawasaki's Theorem*: A single vertex crease pattern can fold flat if and only if the alternating sum of its angles is zero. The current activity shows that this theorem does not extend to multiple vertex crease patterns, as each of the crease patterns presented are made of vertices that satisfy Kawasaki's Theorem. The lower-right crease pattern is especially baffling, since it only contains two vertices! These crease patterns illustrate different ways in which flat foldability can be impossible.

The top row cannot fold flat because they force *mountain-valley contradictions*. This can be seen using a basic fact about flat folding: If we have at a vertex, in order, a large angle then a small angle then another large angle, then the two crease lines in between these three angles must have different mountain-valley parity. The reason for this is that if they were the same, then we'd have two large angles covering a small one on the same side of the paper, which would force the paper to intersect itself. (This is one of the things students may have observed in the Exploring Flat Vertex Folds activity, i.e., the Big-Little-Big Angle Theorem.)

So in the upper-left crease pattern, we have two 90 degree angles surrounding a 60 degree angle at all three of the vertices. Thus, the triangle in the center of the crease pattern is supposed to have mountains and valleys alternating around it, which is impossible.

The upper-right crease pattern suffers from exactly the same thing, but to get the mountain-valley contradiction requires a longer chain of creases. The following picture is one way to do it (Maekawa's Theorem, which states that every flat vertex of degree 4 must have 3 Ms and 1 V or vice versa, is also used):

These examples show that mountain-valley contradictions can be thought of as problems in 2-colorability of graphs. If we look at chains of crease lines that, consecutively, must have different mountain-valley parity, then these chains must be 2-colorable in order to avoid a mountain-valley contradiction. Odd cycles in such chains are the kiss of death.

The bottom two crease patterns in the handout are more difficult to analyze. Neither of them force mountain-valley contradictions. Instead, they force problems with the paper *self-intersecting*. Both are very sensitive to the location of the vertices with respect to the square's boundary. For example, in the bottom-right one, if the two vertices are moved farther apart from each other, then it will be foldable.

Actually proving that the bottom-left crease pattern is impossible is very difficult. Asking students to prove this rigorously is a very good, if somewhat cruel, challenge, and I always do so in hopes that someone might come up with a more solid proof than what I've seen. The idea is that the four corners of the square turn into flaps of paper, and the horizontal and vertical creases surrounding them determine how big these flaps are—the bottom ones are quite large while the top ones are 1/3 the length of the square. All four of these flaps must be wrapped around or tucked inside the model, and if you go through all the possibilities of doing so, you discover that none of them work. Usually the problem is that one of the bigger flaps can remain outside the model, but then the other big flap must be tucked inside, where there isn't enough room. Only experimenting with this model yourself will convince you that this is indeed the case.

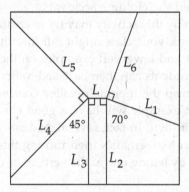

The bottom-right crease pattern on the handout (reproduced above) is the answer to a challenge from the mathematical science writer Barry Cipra to determine whether or not all two-vertex crease patterns that are locally flat-foldable are globally flat-foldable as well. The answer was, "No," and this is an example of such an impossible two-vertex crease pattern. The idea is that the two parallel creases of the 45° and the 70° angles (lines L_2 and L_3) cannot both be valleys or both be mountains, or else the paper will be forced to self-intersect. At the same time, crease lines L_5, L_6, and L have to have the same mountain-valley parity. This is known due to a combination of results: Maekawa's Theorem and the Big-Little-Big

Angle Theorem, which combined tell us that L_3 and L_4 must have different MV parity (thus L and L_5 must be the same) and that L_1 and L_2 must have different MV parity (thus L and L_6 must be the same).

Thus, we can think of the region of the paper formed by L_5, L, and L_6 as being a "back wall" of our fold, in front of which we must arrange the 45° and 70° flaps. One of these flaps can be in front of everything, but the other one will have to be folded inside the model (since L_2 and L_3 are different). No matter which we try to fold inside the model, the L_5-L-L_6 wall will not allow enough room for the fold to lie flat without inducing more creases or ripping the paper.

Class tips

There is a very good chance that you will have students who believe that they have managed to fold one of these crease patterns flat. If this happens, proceed with the confidence that the students must have added an extra crease somewhere or inadvertently moved one of the crease lines. If you have students working on these in groups, you can impose the rule that you won't consider a crease pattern to be successfully folded flat unless everyone in a group is able to duplicate the effort. This will usually catch people who accidentally alter the crease patterns.

The last crease pattern (the lower-right one on the handout), however, is particularly tricky. If students make their creases inaccurately, thus altering the 45° and 70° angles a bit, they may actually get it to fold flat. Therefore it is important to stress that they make the creases as accurately as they can on this model. Folding larger versions of this crease pattern (which can be made by enlarging in on a photocopier) will help avoid such folding accidents.

Students who get taken by this activity may try to create their own impossible crease patterns. If you think your class might fall into this category, try giving them only the upper-right and lower-left examples on the handout. These both have many vertices, and students can then be asked to find examples with fewer vertices. Students who grasp the mountain-valley contradiction concept that is present in the upper-right example will have a good chance at discovering the upper-left example on their own. In fact, several students at the 2005 Hampshire College Summer Studies in Mathematics tried turning this three-vertex example into a two-vertex example by letting one of the vertices be off the paper, as shown below.

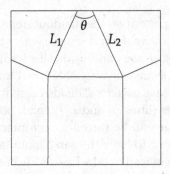

However, this does not quite work. If the angle θ in the above figure is $60°$, as it is in the three-vertex example, then this crease pattern is foldable; merely let L_1 and L_2 have the same mountain-valley parity. If instead, though, we make the angle θ be *less than* $60°$, which will also require that the two vertices be closer together, then the above crease pattern will, indeed, be impossible to fold flat. In this case L_1 and L_2 having the same mountain-valley parity will force the top side of the square to intersect itself when folded. This is similar to, but perhaps easier to comprehend than, the phenomenon encountered in the lower-right two-vertex example on the handout.

Further thoughts and investigations

These examples of impossible crease patterns exhibit problems on the current frontier of flat-foldability research. I published two of these crease patterns (the upper- and lower-left ones on the handout) in a 1994 paper [Hull94], which was also the first paper (that I'm aware of) to communicate the Maekawa and Kawasaki Theorem results. Two years later, Bern and Hayes [Bern96] proved that the general problem of determining whether or not a given crease pattern is flat-foldable is NP-complete. They even proved that if we are *given the mountain-valley assignment* then the problem is still NP! This means that when it comes to deciding flat-foldability, problems with mountain-valley contradictions, like those found in the top two crease patterns on the handout, are easy to detect (can be determined in polynomial time), but the problems of the paper self-intersecting, like in the bottom two crease patterns, are much harder to detect.

This all means that it is the self-intersecting possibilities of paper folding that make modeling origami so difficult. But difficult usually means interesting, as it implies that origami is a lot more complex than one might have originally thought. This is why there are a number of researchers now, like Demaine, Lubiw, and O'Rourke, among others, who are studying the computational complexity problems found in paper folding. Indeed, these researchers, through the papers that they authored in the late 1990s and early twenty-first century, have created a new field of mathematics and theoretical computer science known as *computational origami*. Work in this area has many applications. As mentioned in the Folding and Coloring a Crane activity, no one has managed to make a computer model "virtual origami" perfectly (the NP-completeness mentioned above is a major hinderance), and work in the computational aspects of paper folding would help this effort. There are also applications in robotics and protein folding in biology, as many computational origami problems can be reduced to problems in "one-dimensional folding."

Looking into more advanced problems in computational origami would be a very rich area for undergraduate investigation. Examining the multitude of papers on Erik Demaine's web site (http://theory.lcs.mit.edu/~edemaine) is a great place to start.

Activity 23
FOLDING A SQUARE TWIST

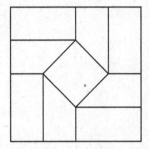

For courses: geometry, discrete math, combinatorics, math for liberal arts, intro to proof, modeling, abstract algebra

Summary

Students are given the crease pattern of a square twist (which can be folded easily enough from scratch too) and challenged to fold it into something flat. When done, the students should compare their models with each other and see if they did the same thing. This leads to a discussion of the difference between mountain and valley creases. Then we ask, "How many different ways can we assign mountains and valleys to the square twist and have it fold up?"

Follow-up activity: What happens when we try folding more than one square twist into the same piece of paper? Is there an organized way in which we can do this?

Content

This is, at heart and when divorced from the other flat folding activities, a modeling problem that involves discrete geometry and combinatorics. The twist fold is also very engaging and requires no overhead, making this a doable activity for a general freshman-level math class. On the other hand, proving one's conclusions in this activity can be tricky to do rigorously, making this a good exercise for students learning proof. Finally, it offers a good situation in which Burnside's Theorem from combinatorial algebra can be applied.

Handout

The handout is written at a general level, where the square twist crease pattern is presented and the basic question of how many different ways can it be folded up is asked. It is left to the students to figure out what exactly they are counting and how to go about doing so.

Time commitment

Folding the square twist will take 25 minutes or so. Depending on how your students choose to enumerate the number of ways to fold it, the rest could take another 20 minutes.

Folding a Square Twist

Activity: Below is shown a crease pattern. The creases are all on the 1/4 lines of the square, but the center diamond needs to be "pinched" in place. Take a square piece of paper and reproduce this crease pattern to see how it folds up.

To help you fold this, follow these instructions:

(1) Fold a 4 × 4 grid of creases on your square.

(2) Pinch the four crease segments that make the diamond in the middle.

(3) Draw the crease pattern below on your creases with a pen.

Then you can try to fold it up.

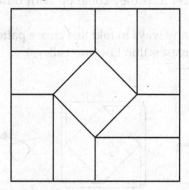

This origami maneuver is called a **square twist** and is one of the less obvious ways in which paper can be folded flat.

Question: Look at your classmates' square twists. Do they look the same as yours? Are you sure? Work together to count how many different ways there are to fold up this crease pattern (without making any new creases).

SOLUTION AND PEDAGOGY

The square twist is a very nontrivial origami move. It represents an intricate way in which we can make a piece of paper "shrink" or contract about a central polygon. (Yes, there are also triangle, hexagon, octagon, etc. twists. You and your students are encouraged to explore those too!)

As such, some students will have a hard time getting this crease pattern to fold into something. I encourage helping students make this crease pattern in a separate square piece of paper, rather than cutting out the one in the handout and just folding that. I suggest this because I feel that actually folding the creases from scratch helps students see this as an interesting property of folded paper, rather than as some weirdly constructed crease pattern. This may help them explore the different possibilities of the crease pattern. Also, keeping the handout whole will allow them to take notes on it as they come up with different ways to fold the square twist.

There are, of course, many ways to fold this crease pattern flat. Below are two (bold lines are mountain creases, thin lines are valleys):

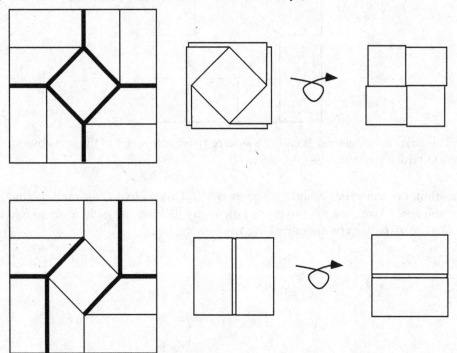

The top example is the "classic" square twist, where it is easy to see how the center diamond rotates by 90 degrees when the crease pattern is folded. This happens in all ways we can fold it flat, although the center diamond is not always completely visible. The bottom example has some nice symmetry properties, in that both sides of the paper are doing the same thing (but rotated). Such models are called *iso-area* by origamists. (See [Mae02] and [Kas87] for more information.)

Now, given any specific way in which a student folds this crease pattern, you can always get another one by switching all the mountain creases to valleys and vice versa. But this quickly brings us to the question of *what exactly are we counting?* Students need to decide if they are counting

(a) physically different folds, or

(b) symmetrically different folds.

In (a) we are really counting the number of valid mountain-valley (MV) assignments as if each crease had a name and so two MV assignments are different if some crease gets a different assignment in them. In (b) we don't want to consider two MV assignments that are the same under rotation to be different.

Both of these problems can be done by sheer exhaustion, since this crease pattern isn't so complex that it forbids going through all the possibilities systematically. In fact, I've had students take this approach, but rarely will they do it properly, delineating their method precisely to prove that no other possibilities exist. Thus, not only is this approach the most lengthy to write up, it's also very hard to get right.

A better approach to (a) is to use some basic facts about flat folds, which the Exploring Flat Vertex Folds activity usually reveals, but which can be independently discovered in this activity. First, notice that each vertex in the square twist crease pattern is the same, with angles between the creases, starting with the inner diamond and going clockwise, 90°, 45°, 90°, and 135°. Maekawa's Theorem ($M - V = \pm 2$, see the Exploring Flat Vertex Folds activity) tells us that at each vertex we must have either three mountains and one valley or vice versa. (Students who haven't done the previous flat folding activities would only need to see that the all four mountains and two mountains, two valleys situations are impossible.) Also, the two crease lines surrounding the 45° angle cannot both be mountains or both be valleys, for otherwise we'd have two 90° angles trying to simultaneously cover a 45° angle on the same side of the paper, which would force the paper to self-intersect. (This was called the "Big-Little-Big Angle Theorem" in the Exploring Flat Vertex Folds activity.)

This all implies that choosing the MV assignment of the inner diamond in the square twist crease pattern will force the rest of the MV assignment. This is because the diamond creases border all the 45° angles, and thus force the crease on the other side of the 45° angle. Then, Maekawa's Theorem forces the remaining crease at each vertex.

Thus the solution to (a) is 2^4 (two choices for each crease in the diamond) or 16 different ways to fold this crease pattern.

If we actually look at all the 16 possible crease patterns, it's easy to see which ones are merely rotations of each other and thus solve part (b). But again, proofs by exhaustion are hard to write up, and there are better tools to use. For example, students could summarize the symmetry in various MV assignments for the inner diamond. This could be done by stating that the diamond can have either four, three, two, one, or no mountain creases (and the rest valley). Breaking it up into these cases and exploring the symmetry of their possibilities can lead to an enumeration of symmetrically-different MV assignments of the square twist.

A more efficient way to do this same thing would be to use Burnside's Theorem (see [Gal01], [Tuc02]), which states that the number of ways N to color an object whose symmetry group is G is

$$N = \frac{1}{|G|} \sum_{\pi \in G} \phi(\pi),$$

where $\phi(\pi) =$ the number of colorings that are fixed under the symmetry π. In our case our group of symmetries is the rotation group of a square, which we will denote $G = \{R_0, R_{90}, R_{180}, R_{270}\}$.

Since R_0 is the identity, we have $\phi(R_0) = 16$.

Since they're inverses of each other, we have $\phi(R_{90}) = \phi(R_{270})$. And if we think about the inner diamond, the only ways in which we could two-color the creases (where our colors are mountain and valley) that would be invariant under $90°$ rotation would be with all mountains or all valleys. Thus $\phi(R_{90}) = \phi(R_{270}) = 2$.

For $180°$ rotations, we could, again, have all mountains or all valleys in our diamond, or we could have them alternate MVMV or VMVM. Thus $\phi(R_{180}) = 4$. So,

$$N = \frac{1}{4}(16 + 2 + 4 + 2) = \frac{24}{4} = 6.$$

The six possibilities are shown below.

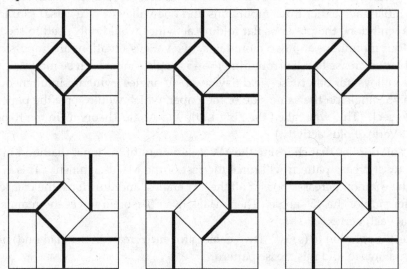

Enterprising students may want to also not count MV assignments that are the same but only with mountains and valleys reversed. In that case the answer would be 4.

Follow-up

Square twists are rather captivating, and you might have students who want to see more origami "like this."

As mentioned previously, other polygons can be "twisted" as well, and trying to twist non-square polygons is possibly the easiest way to expand on this activity. In the first activity in this book, the Folding Equilateral Triangles in a Square activity, one can find instructions for folding a square piece of paper into a regular hexagon. Cutting out such a hexagon will give you an ideal starting point for folding a hexagon twist. The idea is the same as for the square twist; fold a smaller hexagon in the center of the paper, and then fold creases radially out from the corners of your small hexagon. Then make the "pleat" creases parallel to the radial lines.

The below figure shows such a crease pattern, where the smaller hexagon in the center is 1/4 the distance from the edge to the center. Alongside it is an image of what the folded hexagon twist should look like.

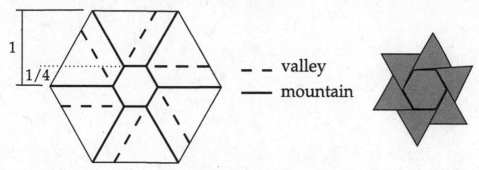

Folding hexagon twists is quite a bit more difficult than square twists, possibly because all six of the "arms" of the twist need to fold at the same time, and so there are simply more creases to manage. But the effect of the twist is rather pleasing, and one can answer the same question: In how many ways can we assign mountains and valleys to this crease pattern in order to fold it flat? (Above is shown only one way; there are many others.)

Another follow-up direction for this activity would be to *tessellate* the square twist into a larger sheet of paper. How to do this is described in Activity 28 (Origami and Homomorphisms).

More information on origami tessellations and making other kinds of twist folds can be found in Eric Gjerde's excellent book *Origami Tessellations* [Gje09].

Activity 24
COUNTING FLAT FOLDS

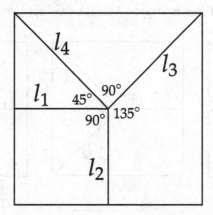

For courses: combinatorics, geometry, discrete math

Summary

Students experiment folding simple, single-vertex flat origami, specifically vertices of degree 4. Students are asked to count the number of ways such vertices can fold flat. From this, bounds on the number of ways that a general degree-$2n$ vertex can fold flat are found.

Content

This activity straddles the fields of combinatorics and geometry. Solving the problems requires basic combinatorial reasoning, but the angles between the creases play a strong role, giving the activity a geometric flavor.

Handout

There is only one handout. It asks students to try folding three different flat vertex folds of degree 4 and to compute how many ways in which they can fold flat. Then two more general questions springboard off these examples.

Time commitment

This will depend greatly on whether students have done some of the other flat-folding activities in this book already. If so, then 20–25 minutes might be enough time for this activity. If not, then expect to give more hints or explicit help on the folding and plan for 40 minutes.

Counting Flat Vertex Folds

Below are shown three different degree-4 origami vertices, v_1, v_2, and v_3.

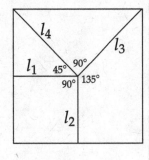 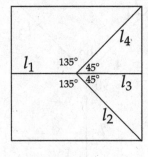 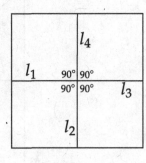

$C(v_1) = $ _____ $C(v_2) = $ _____ $C(v_3) = $ _____

For each of these flat vertex folds, we want to compute

$$C(v) = \text{ the number of ways that } v \text{ can fold flat.}$$

For example, in the third one above, v_3, we could have that l_1, l_2, and l_3 are all valley creases and l_4 is a mountain. That would be one way that v_3 could fold flat.

So fold these vertices using small squares of paper and experiment to compute $C(v)$ for each of them. Then try to answer the following questions.

Question 1: Are there any other values that you think $C(v)$ can take for a degree-4 flat vertex fold than the values you found above?

Question 2: If you had a degree-$2n$ flat vertex fold v, what is the **largest** value that you think $C(v)$ could be? (This would be an *upper bound* on $C(v)$.)

How about the **smallest** (a *lower bound*) value for $C(v)$ that you could get?

SOLUTION AND PEDAGOGY

Degree-4 flat vertex folds are found in almost all flat origami models. Just look at the the crease pattern for the classic crane (see Folding and Coloring a Crane, Activity 20) and you'll see occurrences of all three of the degree-4 folds in this handout.

The basic idea of this activity is so simple that it could be given to any level of math student, at least up to Question 1. However, students who have done some of the other flat-folding activities, like Folding and Coloring a Crane, Exploring Flat Vertex Folds, and Impossible Crease Patterns, will find this activity very straightforward. They might not get a complete solution all the way to the end, but they'll have a much easier time getting started, handling Question 1, and making conjectures for Question 2.

Folding the examples might be nontrivial for your students. Vertex v_3 should not be a problem, even for the most folding-challenged. But for v_1 and v_2, just remember that the crease lines all lie on the easily-folded symmetry lines of the square. To make any of these creases, one needs only to fold half of a diagonal crease or a "fold-in-half book fold" crease, either horizontally or vertically. Making yourself stop at the center and not go all the way across the paper (for creases l_2 and l_4 in v_2 and all of the creases in v_1) might be the hardest part.

Case v_1

Experimentation with this crease pattern should lead students to make some observations about which creases are mountains and which are valleys when v_1 folds flat:

- Creases l_2 and l_3 must always have the same mountain-valley assignment.
- Creases l_1 and l_4 must always have *different* mountain-valley roles.

Students should be encouraged to argue why such observations are true. (Proofs!) The second one is easier to argue using contradiction: Suppose l_1 and l_4 were both valleys, say. (If having them be both mountains were to work, then we could just switch everything and get them both to be valleys, so we may assume without loss of generality that they're both valleys.) Notice that the angle between l_1 and l_4 is 45° whereas the angles between l_1 and l_2 and between l_4 and l_3 are 90°. If l_1 and l_4 were both valleys, then these two 90° angles would both need to cover up the 45° angle, and that's not possible! Either a new crease would have to be created to do this or the paper would have to rip through itself. The only way in which these two 90° angles of the paper can fold around the 45° angle is if l_1 and l_4 have different mountain-valley parity.

This means that there are only two possibilities for the creases l_1 and l_4: either MV or VM. Now, if students have done the Exploring Flat Vertex Folds activity (Activity 21), then they can use Maekawa's Theorem to argue that the creases l_2 and l_3 have to be either MM or VV. (This must be true in order for $|M - V| = 2$ to be true at v_1.) If your students have not done Activity 21, then they will need to argue that l_2 and l_3 must have the same mountain-valley parity (because the angle between l_2 and l_3 is the biggest angle at v_1, and so if these creases were different there is no way that the remaining angles could wrap around the 135° angle to keep the paper from ripping while folding flat).

Therefore, we have two choices for the creases l_1, l_4 and two choices for l_2, l_3, giving us a total of four possible valid MV assignments for v_1. That is, $C(v_1) = 4$.

Case v_2

Notice that the creases must have either three mountains and one valley or vice versa. (Again, this can be deduced by experimentation or by employing Maekawa's Theorem.) Then the key is to notice that l_1 cannot be the lone crease whose mountain-valley assignment is different from the other three. This is because if, for the sake of contradiction, we have that l_1 was a valley and the other creases l_2–l_4 were mountains, then there is no way that the pair of 45° angles would be able to contain the two 135° angles around l_1 and still have it all fold flat.

So assume, for the moment, that we have three mountains and one valley at v_2. Then only l_2, l_3, or l_4 can be the valley, and the rest will be mountains, giving us three possibilities. Each of these can be "flipped" to be a case where we have one mountain and three valleys, giving us that $C(v_2) = 6$.

Case v_3

This is the easiest case. We still must have three mountains and one valley (or vice versa) by Maekawa's Theorem. All of the angles at v_3 are equal, so there are no restrictions as to where, say, the lone valley crease can go and have the rest be mountains. Thus we have four choices for where to put the lone valley crease, and each of these cases could be flipped to become one where we have a lone mountain crease and three valleys. Therefore $C(v_3) = 2 \times 4 = 8$.

In practice, students will get these results by brute-force, trying lots of examples with their paper models of v_1, v_2, and v_3. But it is important that they see some of the logic behind these numbers so that they can extrapolate for the remaining questions.

Question 1

The answer is, "No." Putting together a solid argument as to why this is true would have to use the following observations:

- $C(v)$ is always an even number. This is because for any valid mountain-valley assignment that will fold v flat, we can create another valid mountain-valley assignment simply by reversing the mountain and valley creases. (In the context of Maekawa's Theorem, this establishes a bijection between the MV-assignments that have $M - V = 2$ and those that have $M - V = -2$.)

- For a degree-4 flat vertex fold v, we can't have $C(v) = 2$. $C(v) = 2$ only happens in the abnormal case of a "degree-2" vertex, sitting on a straight line, where either both creases are mountains or both creases are valleys. (We might not want to even consider such degree-2 vertices as existing.) A better argument, though, would be to say that any degree-4 flat vertex fold will have a smallest angle (with possible ties), and on either side of this smallest angle we can either have the creases be MV or VM, and by Maekawa the other creases would have to be the same as each other. (This is the same argument as in the v_1 case above.) That gives us two valid mountain-valley assignments, and each of these can have their mountains and valleys reversed to give us two other valid ways to fold the vertex flat. Thus $C(v) \geq 4$. (Note that this argument will become important for Question 2.)

- The case of vertex v_3 in the handout is clearly the highest value that $C(v)$ can attain for a degree-4 vertex. When all the angles are equal, we have the greatest flexibility for where to put the mountains and valleys; any mountain-valley assignment that preserves Maekawa's Theorem will work! Therefore $C(v) \leq 8$ for vertices v of degree 4. (This argument will also be important for Question 2.)

Thus we have for any degree-4 flat vertex fold v, $4 \leq C(v) \leq 8$, and since $C(v)$ must always be even, the values 4, 6, and 8 are the only ones possible.

Question 2

This question is more sophisticated, asking students to extrapolate what they learned from the folding examples and in Question 1 to a general degree-$2n$ flat vertex fold. (Note that in order for a vertex to fold flat, it needs to have even degree. This is proven in Activity 21, Exploring Fat Vertex Folds, but it's also pretty easy for students to understand why this must be true.)

For the upper bound, students need to realize that the example shown in v_3 generalizes to any vertex. That is, if we want the most number of ways to fold up a vertex, we need to have equal angles between all consecutive creases. But to create a formula for the upper bound, we need to generalize the argument given for case v_3.

It might be easier for students to first consider the degree-6 case with all angles 60° around the vertex. Maekawa's Theorem needs to hold, and so we will have either four mountains and two valleys or vice versa. Suppose it's four mountains and two valleys. Then we can pick any two of the six creases to be valleys and make the rest mountains; there are $\binom{6}{2} = 15$ ways in which to do this. Then any of these can have its mountains and valleys flipped, giving us a total of $15 \times 2 = 30$. So $C(v) \leq 30$ in the degree-6 case.

For a general degree $2n$-vertex, suppose that we have $n + 1$ mountains and $n - 1$ valleys (again using Maekawa). Picking the $n - 1$ valley creases first, and then multiplying by two to flip the mountains and valleys, we get

$$C(v) \leq 2 \binom{2n}{n-1}.$$

For the lower bound, we want to emulate the case v_1, but seeing how this works might be difficult without considering the degree-6 case, again.

For a degree-6 flat vertex fold to have the fewest number of ways to fold flat as possible, we would want to have as little symmetry in the crease pattern as possible. For example, perhaps all of the angles are different. But there will still be a smallest angle, as there is in the v_1 case, and the creases around this smallest angle can either be MV or VM.

Imagine that we fold only the creases around this smallest angle. This will turn our paper into a cone, with the vertex at the apex of the cone. Now, on this cone there will only be four creases left, since we've already folded two of them. Among those four creases there will be a smallest angle, and we can fold its creases using either MV or VM. After that, the remaining two creases must be either MM or VV by Maekawa.

At each juncture in the process that we just described, there were two choices to be made. Therefore the number of ways to fold this vertex would have been $2 \times 2 \times 2 = 8$. So $C(v) \geq 8$ for degree-6 flat vertex folds.

To generalize this argument for an arbitrary degree-$2n$ flat vertex fold, we search for the smallest angle between consecutive creases, fold these two creases in one of two ways (either MV or VM), and then repeat. This will give us at least 2^n ways in which the vertex could have been folded flat, and that is our lower bound.

This argument can be made a lot more rigorous, of course. To do so would require reformulating the conjecture (that 2^n is a lower bound) to be a lower bound for *flat vertex cone folds*, where the piece of paper is a cone with the flat-foldable vertex at the apex. (Then normal paper is just the special case where the cone angle equals 360°.) Then a proper induction argument can be made.

But for the purposes of an in-class activity, simply realizing that we can repeat the process with the smallest angle should be enough for students to give a very convincing argument to each other. Formalizing it with induction would make a good take-home assignment or extra project (and give them practice with induction!). But instructors should not expect students to produce such formalism for flat vertex folds without having already done Activity 21 (Exploring Flat Vertex Folds).

In any case, the full answer to Question 2 is as follows: For a flat vertex fold v of degree $2n$, we have

$$2^n \leq C(v) \leq 2\binom{2n}{n-1}.$$

Further explorations

This activity gives a glimpse at a much harder, research-level question: Given an origami crease pattern that we know can fold flat, in how many different ways can we assign mountains and valleys to the creases to fold it flat? This "counting flat foldings" question is, in general, very difficult.

Almost everything is known about the single-vertex case. The bounds found in this activity are best-possible, since there are examples of flat vertex folds that actually achieve the upper and lower bounds. But for a specific, given flat vertex fold, there exist recursive formulas, based on the angles between the creases, that will give exactly the number of ways it can fold flat. (See [Hull02-1], [Hull03], or [Dem07].)

Another interesting question about the single vertex case is to ask which even-numbered values between the bounds of 2^n and $2\binom{2n}{n-1}$ can be attained by a degree-$2n$ flat vertex fold. In the degree-4 case, all even values between 4 and 8 can be realized, but in the degree-6 case *not* all values between 8 and 30 can be achieved. Figuring out exactly which numbers $C(v)$ can be for a degree-6 vertex is a very fun and challenging exercise for those who enjoy folding lots of examples and carefully documenting them. For a general degree-$2n$ vertex, however, this problem is open. Progress on this problem, at least as of 2011, can be found in a paper by myself and my former student Eric Chang [HullCha11].

For crease patterns with more than one vertex, the game is more-or-less wide open. Only specific examples of crease pattern families have been looked at, and even the most simple of cases seem to be quite complex. For example, there is the postage stamp-folding problem, which basically is considering crease patterns that merely form an evenly-spaced $m \times n$ grid on the paper. Quite a bit of work has been done on this problem, and while researchers have been able to compute, using careful algorithms, the number of ways to fold such grids flat, anything like closed formulas remain elusive. (See [Koe68] and [Lun68].)

It is also noteworthy that such counting folding problems are of interest in physics and physical chemistry. Specifically, people who study how polymer membranes crumple have to face this problem. Polymer membranes, whether man-made or naturally occurring, such as in the walls of blood cells in the human body, are made of some kind of molecular lattice, such as a square or triangle lattice. If such a polymer were to crumple, then it would literally fold along the molecular bonds that make up this lattice. (It might not use *all* the bond-creases, but it will use some of them.) In order to understand the mechanical properties of such polymers, knowing approximately how many ways in which they can fold or crumple up is key, and physicists will employ methods of thermodynamics to attain such approximations (because when such folds are made, energy is being released, so thermodynamics can provide a model for what is going on). The mathematics involved in such study is pretty intense and quite different from the methods explored in this activity, but interested readers or students should consult Philippe Di Francesco's survey paper, "Folding and coloring problems in mathematics and physics" [DiF00].

Activity 25
SELF-SIMILAR WAVE

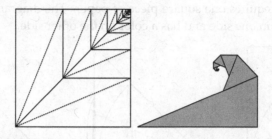

For courses: fractal geometry, complex analysis

Summary

Instructions are passed out to make an origami wave. Once the class has made the wave, students are asked to examine its spiral nature. If we put the model in a set of coordinate axes, can we find the point at which the model's spiral converges?

Content

This is a single-sheet model that makes an origami wave. It is an example of a self-similar origami model, and the math behind it can be explored using either transformational geometry or complex numbers. This wave is also an example in a genre of origami known as "infinite progression" folds, and often people mistakenly identify such models as being fractals. Therefore, for a class studying fractals, this model offers a good, hands-on exploration of an object that exhibits self-similarity, but that is *not* a fractal.

Handout

The handout simply provides the instructions for making this wave model. Students are left to use the techniques they've been learning in class (either geometric transformations or complex numbers) to solve the puzzle.

Time commitment

This model is one in which the crease pattern must be folded first, and then the whole model can be folded. Also, it is up to the students or the instructor to decide how many "levels" they want to fold into the model. Folding fewer levels, like only 3 or 4, will take less time, probably 15–20 minutes.

The math side of this activity is pretty involved, requiring a strong background in simple algebra and/or complex numbers to see it to the end. Groups of students will generally need 20–30 minutes to complete the activity, although this depends on their background. Planning a whole hour for the activity is a safe bet.

The Self-Similar Wave

This wave model requires one square piece of paper. The diagrams assume that the paper is white on one side and has a color on the other side.

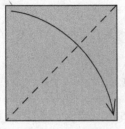

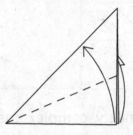

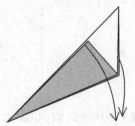

(1) Looking at the color side, fold a diagonal.

(2) Fold one layer up to the diagonal. Repeat behind.

(3) Unfold step (2).

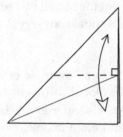

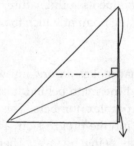

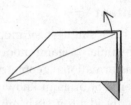

(4) Fold perpendicularly to the right side at the indicated spot.

(5) Now use the creases from (4) to reverse the point inside...

(6) ...like this. Crease sharply and unfold step (5).

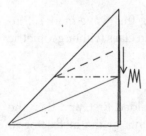

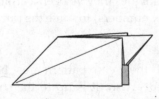

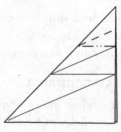

(7) Refold step (5), but this time add angle bisectors with the diagonal to crimp the paper.

(8) This should be the result. Crease firmly and unfold.

(9) Now repeat steps (4)–(8) to make the next "level."

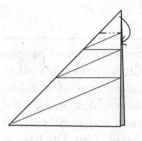
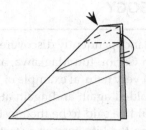
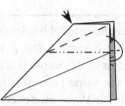

(10) You could keep going, but for the first time, stop after 3 levels by performing steps (4)–(5) one last time.

(11) Then use the creases of the third level to swivel the paper inside.

(12) Do it again with the second level creases. The wave spiral will be forming inside.

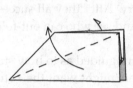
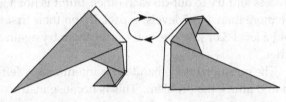

(13) For the first level, all you need to do is refold the creases from step (2).

(14) This reveals the wave! Of course, you can, and should, do more levels to make the wave curl more.

Question: Suppose we started with a square piece of paper with side length 1 and folded this wave with an infinite number of levels. If we put the finished model on a set of coordinate axes, with the tip of the base at the origin, as shown below right, what would the coordinates of the limit point P of the spiral be?

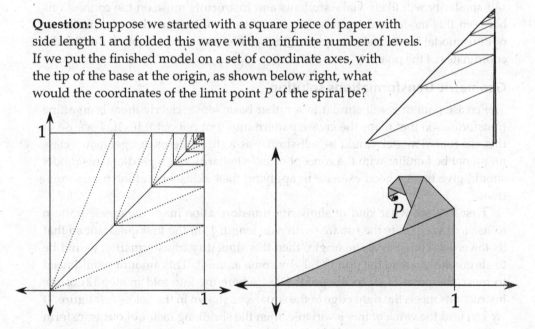

SOLUTION AND PEDAGOGY

This origami model has been independently discovered by many different people, including Paulo Baretto, Ilan Garibi, Jun Maekawa, and Chris Palmer, as well as the author of this book. It is a very natural example of "infinite progression" folds, where the same pattern is folded again and again at smaller scales. In fact, the Self-Similar Wave model could be said to be the most natural way to do such an infinite progression fold using the kite base as a starting point. (The kite base is basically the first two steps of the model.)

Folding this model has some tricky steps, and instructors should practice it themselves a few times before asking a whole class to try it. One way to make sure that your students have success folding this model is to tell them to only fold 3 levels of the iteration on their first try. You may have to be firm about that, since students will quickly see how they could just keep "folding to infinity" in this process and try to out-do each other. But it is not a good idea for novice folders to try more than 4 or 5 levels, especially on their first try. After they all successfully fold a level-3 or -4 wave, you can let them try again and see who can out-fold each other.

The question on the handout is intentionally left unguided as to how students should attack the problem. This is because instructors might want their students to use a specific approach related to their class. I know of two approaches that lend themselves to curricula in standard college-level (or advanced high-school) classes: using geometric transformations or complex numbers. Both solutions take advantage of the self-similarity of the model and the crease pattern, and in fact this self-similarity will likely make students and instructors muse on the connections between this model and fractals. We will return to this question of whether or not this model is a fractal after presenting solutions to the question of finding the coordinates of the point P.

Geometric transformations solution

The crease pattern is self-similar in a rather basic sense; clearly there is an affine transformation that maps the crease pattern into (but not onto) itself. Because of this, the folded model should be self-similar as well. Students in a geometry class might not be familiar with the concept of self-similarity, but regardless this model should give them a good exercise in applying their skills with affine transformations.

First let's see what kind of similarity transformation maps the crease pattern to itself. If we situate the square (with side length 1) in the first quadrant so that its lower-left corner is at the origin, then this similarity transformation would be to shrink the plane to the point $(1, 1)$ by some amount. This amount would need to move the point $(1, 0)$ to the point $(1, y)$ where the kite fold in step (2) of the instructions meets the right edge of the square, as shown in the below-left figure. If we can find the value of this y variable, then the shrinking factor of our transform will be $1 - y$, which is the side length of our smaller square under the similarity transformation.

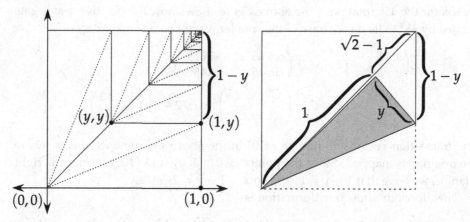

The value of y can be found in a number of different ways. The above right figure illustrates one way, where it is shown how $1 - y$ is the hypothenuse of a 45° right triangle where one leg has length $\sqrt{2} - 1$ and the other has length y. But since the legs of a 45° right triangle have the same length, this means $y = \sqrt{2} - 1$.

And so the shrinking factor of our similarity transformation is $1 - y = 2 - \sqrt{2}$.

This shrinking factor will be the same for the similarity transformation of the folded wave. It can be difficult to see exactly how the self-similarity of the wave works, even if one has a folded model in their hands. Below is an illustration of what our folded wave looks like if we folded it with translucent paper, revealing the creases inside.

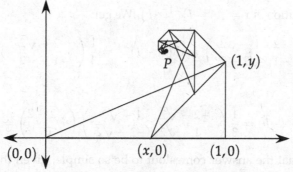

This shows how the point $(1, 0)$ will still map to the point $(1, y)$ on the folded wave, and the origin will map to a point $(x, 0)$. In fact, it also illustrates how the point P at the center of the wave's spiral will be a *fixed point* of our affine transformation. That is the key; if we can find a formula for this affine transformation, then P will be the unique fixed point of the transformation.

Our transformation will be a function $F(x, y)$ of the form

$$F(x,y) = \begin{pmatrix} a & b \\ c & d \end{pmatrix} \begin{pmatrix} x \\ y \end{pmatrix} + \begin{pmatrix} e \\ f \end{pmatrix}$$

where the matrix $\begin{pmatrix} a & b \\ c & d \end{pmatrix}$ is a shrinking and a rotation, and the vector $\begin{pmatrix} e \\ f \end{pmatrix}$ is a translation vector. The shrinking will be $2 - \sqrt{2}$, and we can use the standard rotation matrix for the 45° rotation. (The above figure shows how the positive x-axis gets rotated by 45° to lie on the base of the smaller wave.) So our matrix is

$$\begin{pmatrix} a & b \\ c & d \end{pmatrix} = (2 - \sqrt{2}) \begin{pmatrix} \cos 45° & -\sin 45° \\ \sin 45° & \cos 45° \end{pmatrix}$$

$$= (2 - \sqrt{2}) \begin{pmatrix} \frac{\sqrt{2}}{2} & -\frac{\sqrt{2}}{2} \\ \frac{\sqrt{2}}{2} & \frac{\sqrt{2}}{2} \end{pmatrix} = (\sqrt{2} - 1) \begin{pmatrix} 1 & -1 \\ 1 & 1 \end{pmatrix}.$$

The translation vector will just be $(x, 0)$ in the above figure, since this is where the origin gets mapped. Since the points $(x, 0)$, $(1, y)$, and $(1, 0)$ form a 45° right triangle, we have that $1 - x = y$, and so $x = 1 - y = 2 - \sqrt{2}$.

Therefore our affine transformation is

$$F(x, y) = (\sqrt{2} - 1) \begin{pmatrix} 1 & -1 \\ 1 & 1 \end{pmatrix} \begin{pmatrix} x \\ y \end{pmatrix} + \begin{pmatrix} 2 - \sqrt{2} \\ 0 \end{pmatrix}.$$

Now, one could solve the equation $F(x, y) = (x, y)$ using nothing but algebra, which is a good exercise in multiplying by square-root conjugates to rationalize denominators and such. If you or your students like that sort of thing, then hop to it.

But the system $F(x, y) = (x, y)$ can be solved more readily using matrices. If we think of it as a matrix-vector equation $A\vec{x} + \vec{b} = \vec{x}$, then we can rewrite it as

$$A\vec{x} - \vec{x} = -\vec{b} \;\Rightarrow\; (A - I)\vec{x} = -\vec{b}$$

and thus our solution is $\vec{x} = (A - I)^{-1}(-\vec{b})$. We get

$$A - I = \begin{pmatrix} \sqrt{2} - 2 & 1 - \sqrt{2} \\ \sqrt{2} - 1 & \sqrt{2} - 2 \end{pmatrix} \text{ and } (A - I)^{-1} = \frac{1}{3} \begin{pmatrix} -2 - \sqrt{2} & 1 + \sqrt{2} \\ -1 - \sqrt{2} & -2 - \sqrt{2} \end{pmatrix}.$$

So our solution is

$$P = (A - I)^{-1}(-\vec{b}) = \frac{1}{3} \begin{pmatrix} -2 - \sqrt{2} & 1 + \sqrt{2} \\ -1 - \sqrt{2} & -2 - \sqrt{2} \end{pmatrix} \begin{pmatrix} \sqrt{2} - 2 \\ 0 \end{pmatrix} = \begin{pmatrix} 2/3 \\ \sqrt{2}/3 \end{pmatrix}.$$

It is surprising that the answer comes out to be so simple, given the potential to get messy with lots of $\sqrt{2}$s. Yet this solution does offer some solid practice with affine transformations.

Complex numbers solution

If instead we consider our folded wave to be sitting in the complex plane, we can get a solution to the location of the point P by writing it as the infinite sum of a well-chosen sequence of points P_n on the folded model. There are several ways in which this could be done, but perhaps the most simple is to follow the image of the main diagonal of the original square in the folded wave. This is illustrated in the following figure.

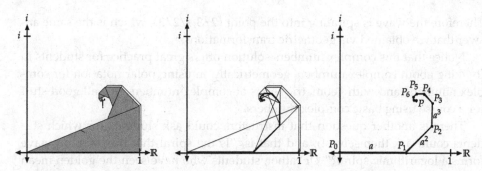

The first segment of our path, from P_0 to P_1 has length a. On the unfolded square this corresponds to the first section of the main diagonal crease (the one from the origin to the point (y, y) in a previous figure). The work we did in the geometric transformations solution shows that $a = 2 - \sqrt{2}$, so the first point on our P_n path is $P_1 = 2 - \sqrt{2}$ in the complex plane.

To go from P_1 to P_2, we need to rotate up by $45° = \pi/4$ radians and go a distance of a^2. If we write our complex numbers in the form $re^{i\theta}$, then computing the location of P_2 is easy:

$$P_2 = P_1 + a^2 e^{\frac{\pi}{4}i} = a + a^2 e^{\frac{\pi}{4}i}.$$

Then for P_3 we travel to P_2 and then swing an angle of $2(\pi/4)$ (measured from the positive real axis) and travel a length of a^3. So

$$P_3 = P_2 + a^3 e^{2(\pi/4)i} = a + a^2 e^{(\pi/4)i} + a^3 e^{2(\pi/4)i},$$

and continuing in this way we get

$$P_n = a + a^2 e^{\frac{\pi}{4}i} + a^3 e^{2\frac{\pi}{4}i} + \cdots + a^n e^{(n-1)\frac{\pi}{4}i}.$$

So the point P will be the infinite sum

$$P = a \sum_{n=0}^{\infty} (a e^{\frac{\pi}{4}i})^n.$$

But this is just a geometric sum: $\sum_{n=0}^{\infty} z^n = 1/(1-z)$ as long as $|z| < 1$. Here $z = a e^{(\pi/4)i} = (2 - \sqrt{2})(\cos(\pi/4) + i\sin(\pi/4)) = (2 - \sqrt{2})((\sqrt{2}/2) + (\sqrt{2}/2)i) = (\sqrt{2} - 1)(1 + i)$. So we get (using the fact that $a = 2 - \sqrt{2} = 2/(2 + \sqrt{2})$)

$$P = (2 - \sqrt{2}) \frac{1}{1 - (\sqrt{2} - 1)(1 + i)} = \frac{2}{2 + \sqrt{2}} \frac{1}{(2 - \sqrt{2}) - (\sqrt{2} - 1)i} = \frac{2}{2 - \sqrt{2}i}.$$

Multiplying the top and bottom of this fraction by $2 + \sqrt{2}i$ yields

$$P = \frac{2}{2 - \sqrt{2}i} \frac{2 + \sqrt{2}i}{2 + \sqrt{2}i} = \frac{4 + 2\sqrt{2}i}{6} = \frac{2}{3} + \frac{\sqrt{2}}{3}i.$$

Therefore the wave is spiraling into the point $(2/3, \sqrt{2}/3)$, which is the same answer that we obtained via geometric transformations.

Notice that this complex numbers solution offers great practice for students in thinking about complex numbers geometrically, in using polar notation for complex numbers, and with geometric series of complex numbers. It's all good stuff for a course using basic complex numbers.

There is another question that instructors could ask students (or which students could ask themselves!), and that is, "Is the spiral that this origami wave forms a logarithmic spiral?" Or rather, students who have seen the golden mean spiral might wonder if the Self-Similar Wave spiral is related.

A spiral is *logarithmic* if the distance between points on the spiral and the center point P increases exponentially as we travel out on the spiral. (As opposed to *Archimedean spirals* that move away from their center at a constant rate. The golden mean spiral is an example of a logarithmic spiral.) Thus if our self-similar wave is modeling a logarithmic spiral, the distance $|P - P_n|$ should be an exponential function in n. Let's see if it is:

$$|P - P_n| = \left| a \sum_{k=n}^{\infty} (ae^{\frac{\pi}{4}i})^k \right| = \left| a(ae^{\frac{\pi}{4}i})^n \sum_{k=0}^{\infty} (ae^{\frac{\pi}{4}i})^k \right|$$

$$= \left| (2 - \sqrt{2})((\sqrt{2} - 1)(1 + i))^n \right| \cdot |P|$$

$$= (2 - \sqrt{2})((\sqrt{2} - 1)\sqrt{2})^n \frac{\sqrt{6}}{3} = \frac{\sqrt{6}}{3}(2 - \sqrt{2})^{n+1}.$$

(Here we used the fact that $|1 + i|$ is the distance from the origin to the point $1 + i$, which is $\sqrt{2}$.) We see that the distance between P and the points P_n on the spiral is an exponential function. Therefore the spiral that the self-similar wave produces is, indeed, a logarithmic spiral. If we translate this spiral so that P is at the origin, we can obtain the polar coordinate equation for this spiral:

$$r(\theta) = \frac{\sqrt{6}}{3}(2 - \sqrt{2})^{\frac{4}{\pi}(\theta - \pi - \arctan(\sqrt{2}/2))}.$$

A graph of this spiral is shown below.

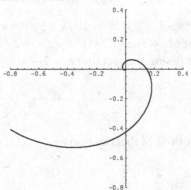

Connection with fractals

As mentioned previously, the Self-Similar Wave model is only one in a large family of origami crease patterns that infinitely repeat themselves and yet are foldable. Two more examples of such crease patterns are shown below. The one on the left was devised by Jun Maekawa, and on the right is a crease pattern due to Shuzo Fujimoto.

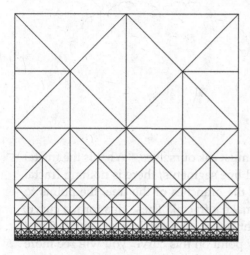 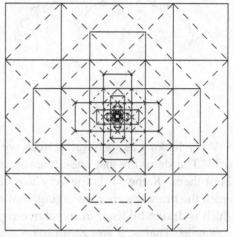

These two crease patterns are *not easy to fold*! They are substantially more difficult than the Self-Similar Wave. They are merely presented here as more examples of origami crease patterns that exhibit self-similarity.

There is a tendency of origamists to refer to such origami models as "origami fractals." Students who have some exposure to fractals might try to say this as well. However, these origami models are not really exhibiting fractal behavior, and calling them fractals is a mistake.

Because fractal geometry is still a relatively new area of math, there is still some debate over what the technically precise definition of "fractal" is. However, the basic, working definition of a fractal is an object whose Hausdorff dimension is strictly greater than its topological dimension. Here *topological dimension* is the standard concept of dimension that students know, where a point is zero-dimensional, a line or smooth curve is one-dimensional, a set in the plane with area is two-dimensional, and so on. *Hausdorff dimension* is very technical, but it can be shown to be the same as what many books call "similarity dimension" or "fractal dimension" where the self-similar nature of an object is related to its dimension. This concept of dimension can end up being non-integer, and so most fractals have Hausdorff dimensions equalling non-whole numbers, like 1.26 or 0.792.

Another way to think about this is that a fractal is an object that exhibits some type of self-similarity *at all levels*. That is, as you zoom in on any point of the object, you will continue to see more copies of its self-similar nature.

The crease pattern and folded model of the Self-Similar Wave does seem like it would satisfy this latter, intuitive definition. But comparing it to an actual fractal reveals the problem with such thinking. Below is shown an example of a fractal tree. The angle between the branches is the same throughout the tree, and the lengths of the stems of the tree decrease at an exponential rate as the number of stems increases exponentially.

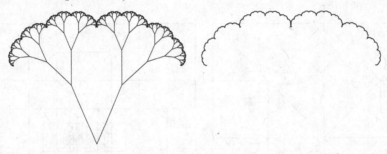

When looking as such fractal trees, we must ask ourselves, "What is the fractal part?" The stems are just one-dimensional line segments; there is nothing fractal about them. Rather, it is the very tips of the branches of the tree, the points to which the branches are converging, that make a fractal. These tips form a curve, which is drawn by itself in the above-right figure, that exhibits self-similarity *at every point*. That is, if you zoom-in on any point on this curve, you will see more and more bumps and copies of the curve. In contrast, if you zoom-in on one of the stems of the tree, you'll just see a straight line, not more and more copies of the tree.

It is this question, "What is the fractal part?" that we must also ask of our Self-Similar Wave origami model. Most of the wave's crease pattern acts like the stems of the fractal tree; they do not make up the "infinite part." In fact, the self-similar nature of the crease pattern, and of the folded wave, is converging to a single point. In the crease pattern this point is the upper-right corner of the square, and in the folded wave it is the point P. In both cases, the "infinite part" is just a single point, which has dimension zero. This is not a fractal.

In the other two self-similar crease patterns shown earlier, the same kind of thing happens. Jun Maekawa's example has the crease pattern converging to a straight line at the bottom of the square, which has dimension one. In Shuzo Fujimoto's example, the creases converge to the center point of the square, which has dimension zero. Thus, neither of these are a fractal.

Folding a piece of paper to mimic an actual fractal is very difficult. Japanese origamist Ushio Ikegami has managed to successfully design and fold an origami fractal tree. (See [Ike09].) This model by Ikegami is extremely difficult to fold, however, and serves as a testament to how difficult it is to fold an actual fractal.

Activity 26
MATRIX MODEL OF FLAT VERTEX FOLDS

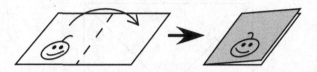

For courses: geometry, linear algebra, modeling

Summary

This activity takes the following approach to modeling paper folding: When we fold a piece of paper flat, we're really reflecting one part of the paper onto the other. Thus, every time we make a flat fold, we're performing a reflection. Reflections of the plane can be modeled with matrices. So, students are given a simple, four-valent flat vertex fold and asked to compute the 2×2 reflection matrices for each of the crease lines. Then, they are asked what they get when they multiply these matrices together. Does it make sense that we get the identity?

Content

This is an application of linear algebra, although the connection to geometry makes this suitable to either a linear algebra or geometry course where basic matrix operations can be assumed. The main result of this activity, that the product of the reflections about creases, in order, about a vertex will be the identity if and only if the vertex folds flat, is actually equivalent to Kawasaki's Theorem (from the Exploring Flat Vertex Folds activity).

Handout

The handout is self-explanatory, leading the students through the activity of generating the folding matrices and challenging them to discover what happens when we multiply them.

Time commitment

Depending on how good your students are at constructing reflection matrices, this activity could be fast, taking only 15–20 minutes, or longer, taking 30–40 minutes.

Matrices and Flat Origami

Idea: When we fold a piece of paper flat, we're really **reflecting** one half of the paper onto the other half. We can use this to model flat origami using matrices.

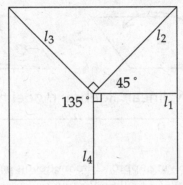

Activity: Above is shown the creases of a flat vertex fold. Assume that the vertex is located at the origin of the xy-plane.

Question 1: Find a 2×2 matrix $R(l_1)$ that reflects the plane about crease line l_1. Do the same thing for the other crease lines.

Question 2: What happens when you multiply these matrices together? Explain what's going on.

SOLUTION AND PEDAGOGY

This activity is about flat origami, which encompasses all origami in which the final model can be pressed in a book without crumpling or adding new creases. The previous activities Exploring Flat Vertex Folds, Impossible Crease Patterns, and Folding a Square Twist provide a good introduction to this topic. While students don't need to have seen these previous activities to engage in this one, instructors may find perusing these previous activities very useful.

Before the students begin producing matrices, give them small squares of paper and have them fold the vertex shown on the handout. Each crease line should be made separately; l_1 and l_4 are made by folding the paper in half from side to side, but not creasing all the way through (stopping at the center), and l_2 and l_3 are made by folding diagonals of the square (again, stopping at the center). Then, all creases should be folded at the same time (say, l_1 a mountain and l_2–l_4 valleys) to obtain a flat vertex fold. Having a model in hand to look at will get the idea of flat folding across to students and help them visualize the reflection matrices that they'll need to produce.

Students in geometry or linear algebra classes who have recently played with matrices of various isometries of the plane should have no problem with the first part of the activity. Sometimes students have a hard time with reflecting about the line $y = x$ or $y = -x$. For such students suggestions can be made on how to figure this kind of thing out. For example, reflecting about $y = x$, which is l_2, should send the point $(1, 0)$ to $(0, 1)$ and the point $(0, -1)$ to $(-1, 0)$. So our unknown 2×2 transformation matrix (with entries $a, b, c,$ and d) should satisfy

$$\begin{pmatrix} a & b \\ c & d \end{pmatrix} \begin{pmatrix} 1 \\ 0 \end{pmatrix} = \begin{pmatrix} 0 \\ 1 \end{pmatrix} \text{ and } \begin{pmatrix} a & b \\ c & d \end{pmatrix} \begin{pmatrix} 0 \\ -1 \end{pmatrix} = \begin{pmatrix} -1 \\ 0 \end{pmatrix}.$$

Staring at this for long enough can allow students to figure out what the variables are. Or they can multiply them out, get four equations in four unknowns, and solve.

In any case, if $R(l_i)$ is the reflection matrix about crease l_i, then the solution to Question 1 is

$$R(l_1) = \begin{pmatrix} 1 & 0 \\ 0 & -1 \end{pmatrix},$$

$$R(l_2) = \begin{pmatrix} 0 & 1 \\ 1 & 0 \end{pmatrix},$$

$$R(l_3) = \begin{pmatrix} 0 & -1 \\ -1 & 0 \end{pmatrix},$$

$$R(l_4) = \begin{pmatrix} -1 & 0 \\ 0 & 1 \end{pmatrix}.$$

In Question 2, multiplying these matrices together should give

$$R(l_4)R(l_3)R(l_2)R(l_1) = I.$$

Or one could put the matrices in the reverse order. But the multiplication should occur in the same order in which we encounter the crease lines, either going clockwise or counterclockwise about the vertex.

One partial reason for why we get the identity matrix is fairly simple: Multiplying the matrices in order is simulating the orientation of a bug walking around the vertex on the flat-folded model. Since the bug should come back to where it started, in the same orientation, the product of those matrices should be the identity.

However, while this argument is leading in the right direction, and can be made to work with care, when stated this simply it is seriously flawed. Therefore, we will present multiple proofs for the result of Question 2.

Proof 1: Formalizing the bug-walking.

Suppose that we let the region between crease lines l_1 and l_4 (the lower-right quadrant) be fixed and fold the rest of the regions according to the crease lines. Let our bug begin in the fixed region and follow a path that goes counterclockwise about the vertex on the unfolded paper (but our bug will be walking on the folded paper, remember).

The bug will first walk across crease l_1, and the reflection that the bug will make will be $R(l_1)$. Fine. But then the bug will continue to walk and eventually encounter crease l_2, except l_2 will no longer be in the position it was on the unfolded sheet. So the reflection matrix that models the bug walking around this second crease will *not* be $R(l_2)$! It will be whatever the reflection is about the *image* of l_2 after the folding is done. Call this matrix L_2. Then the bug will continue walking and reflect about the images of l_3 and l_4 after they are folded; call these matrices L_3 and L_4. Then, since the bug returns to the same region where it began, we should have

$$L_4 L_3 L_2 L_1 = I$$

where we write $L_1 = R(l_1)$ to make it look nice. *This* is what students are likely to get if they try a straightforward, bug-walking approach, but it is not the same thing as the product of the $R(l_i)$ matrices.

Nonetheless, this is a good direction in which to proceed. Let us compute the matrix L_2. One way to model the operation of this reflection is to first *unfold* crease l_1, then do $R(l_2)$, then refold l_1. Thus, we get

$$L_2 = L_1 R(l_2) L_1^{-1} = R(l_1) R(l_2) R(l_1)^{-1}.$$

Similarly, L_3 can be modeled by unfolding l_2 (in folded position), unfolding l_1, then performing $R(l_3)$ and refolding l_1 and l_2 (in folded position). Thus,

$$\begin{aligned}
L_3 &= L_2 L_1 R(l_3) L_1^{-1} L_2^{-1} \\
&= (R(l_1) R(l_2) R(l_1)^{-1})(R(l_1)) R(l_3)(R(l_1)^{-1})(R(l_1) R(l_2)^{-1} R(l_1)^{-1}) \\
&= R(l_1) R(l_2) R(l_3) R(l_2)^{-1} R(l_1)^{-1}.
\end{aligned}$$

Similarly,

$$L_4 = L_3 L_2 L_1 R(l_4) L_1^{-1} L_2^{-1} L_3^{-1}$$
$$= R(l_1) R(l_2) R(l_3) R(l_4) R(l_3)^{-1} R(l_2)^{-1} R(l_1)^{-1}.$$

Then notice that

$$I = L_4 L_3 L_2 L_1$$
$$= (R(l_1) R(l_2) R(l_3) R(l_4) R(l_3)^{-1} R(l_2)^{-1} R(l_1)^{-1})$$
$$\cdot (R(l_1) R(l_2) R(l_3) R(l_2)^{-1} R(l_1)^{-1})$$
$$\cdot (R(l_1) R(l_2) R(l_1)^{-1}) \cdot (R(l_1))$$
$$= R(l_1) R(l_2) R(l_3) R(l_4).$$

You can see how this could be generalized to flat vertex folds of any degree. But this is not for the faint of heart when it comes to symbolic matrix manipulation! Thus, pursuing this line of argument in full detail is not the easiest proof for students to construct or follow. It is a great exercise in linear algebra and geometric transformations, though.

Notice, though, how this proof highlights how surprising it is that the product of the $R(l_i)$ gives us the identity. Remember that these matrices are all reflecting about crease lines *in their original positions*! The crawling bug argument makes perfect logical sense, but that is *not* what $\prod R(l_i)$ is doing.

Proof 2: A bug variation. A careful approach can make a variation of the bug-crawling argument lead to a more simple proof. Let F be the region of the paper between crease lines l_1 and l_4. Imagine that we rip this region in two, tearing the paper from the boundary of the square (say, the point $(1, -1)$) to the origin. This turns F into two smaller regions: F', which is adjacent to l_1, and F'', which is adjacent to l_4. Now perform our reflections $R(l_4)$ through $R(l_1)$ in sequence to the region F''. Each reflection will be simulating what the paper is doing as the creases are folded, and since the vertex folds flat in the end, we must have F' and F'' lining up along their tear. That is,

$$R(l_1) R(l_2) R(l_3) R(l_4)[F''] = I.$$

This only proves that this matrix product is the identity on the region F''. Astute students of linear algebra may understand, however, that in this case it implies that the matrix product will be the identity for *all* points in the plane, not just those in F''.

The reason for this is because F'' is a region of the xy-plane with positive area, and so there exist two linearly independent vectors in F'' (when positioned at the origin); call them v_1 and v_2. Also, for convenience let us denote the product $R(l_1) R(l_2) R(l_3) R(l_4)$ by the matrix T. So we have that

$$T(v_1) = v_1 \text{ and } T(v_2) = v_2.$$

Now let v be any other vector in the plane. Then $v = av_1 + bv_2$ for some scalars $a, b \in \mathbb{R}$. Now, T is a linear transformation, so we have

$$T(v) = T(av_1 + bv_2) = aT(v_1) + bT(v_2) = av_1 + bv_2 = v.$$

Therefore T is the identity matrix.

Please note that the last part of the argument presented here is not really necessary. T is an isometry, and so it is uniquely determined by its action on three non-collinear points. Since T leaves the origin and the endpoints of v_1 and v_2 (when positioned at the origin) fixed, we have that T must be the identity.

But seeing that this follows directly from the fact that T is a linear transformation can be a very valuable experience for students in a linear algebra class. It provides a concrete application of the linear transformation definition, which can often feel abstract to students.

It is also worth noting that this trick of using the fact that T is a linear transformation to prove that T acts like the identity will *not* work in the next activity, where we will examine non-flat, 3D folds. (This is because the three point determination theorem for isometries does not hold in $\mathbb{R}^3$, but we will leave such a discussion for Activity 27.)

Proof 3: Use Kawasaki's Theorem! If students shrink from detailed linear algebra, this result can be proven rigorously for any flat vertex fold using a version of Kawasaki's Theorem. This might have been encountered by students in the Exploring Flat Vertex Folds activity. It states that if $\alpha_1, \alpha_2, \cdots \alpha_{2n}$ are the angles, in order, between the creases at a vertex, then this vertex will fold flat if and only if $\alpha_1 + \alpha_3 + \cdots + \alpha_{2n-1} = 180°$ and $\alpha_2 + \alpha_4 + \cdots \alpha_{2n} = 180°$. (Or alternatively, that $\alpha_1 - \alpha_2 + \alpha_3 - \alpha_4 + \cdots - \alpha_{2n} = 0$.)

The idea is to use the fact that the product of two reflections is a rotation, and the rotation will be twice the angle between the two reflection lines. Thus, when we do $R(l_2)R(l_1)$ and α_1 is the angle between the creases l_1 and l_2, we get a rotation by $2\alpha_1$. Then, $R(l_4)R(l_3)$ will be a rotation by $2\alpha_3$. Continuing in this way, we get that our product of reflection matrices is a rotation of the plane by angle

$$2\alpha_1 + 2\alpha_3 + \cdots + 2\alpha_{2n-1}.$$

By Kawasaki's Theorem, this equals 360°, and thus the product of the reflections is the identity matrix.

The converse can be proven in the same way: If the product of the reflection matrices generated by the crease lines of a vertex fold is the identity, then the creases can fold flat (assuming the converse direction of Kawasaki).

In the theory of flat origami, this model is very useful. One can extend the basic result of this exercise for general, multiple-vertex crease patterns in the following way: Let γ be any vertex-avoiding closed curve on the crease pattern of a flat origami model, and let $R(\gamma)$ denote the product of the reflections about the crease lines that γ crosses, in order. Then $R(\gamma) = I$. (See [bel02] for details.) This is a necessary condition for general flat origami crease patterns, but it is not a sufficient condition. See the Impossible Crease Patterns activity for examples.

Also, this model tells us a lot about what the paper does when it folds flat. If we take a face F of a flat origami crease pattern and decide that F will remain fixed as we fold the rest of the paper, then we can define the *folding map* to be the image of any other face F' in the crease pattern under reflections $R(\gamma)$, where here γ is a vertex avoiding *path* from a point in F to a point in F'. It can be shown that this map is well-defined, and it tells us where each region of the paper goes when the paper is folded flat. For more information, see [bel02] and [Jus97].

Activity 27
MATRIX MODEL OF 3D VERTEX FOLDS

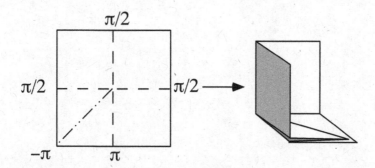

For courses: geometry, linear algebra, modeling

Summary

This is really a follow up on the Matrix Models of Flat Vertex Folds activity. The concept is the same: The product of rotation matrices, in some sense, around a three-dimensional vertex fold should give us the identity. But being in three dimensions, the rotation matrices are more challenging, and it's more complicated to prove that the product of the crease pattern matrices, in the proper order, will return the identity. (We cannot rely on Kawasaki's Theorem here!)

Content

This is a very challenging linear algebra application to three-dimensional geometry. It requires a solid command of rotations in $\mathbb{R}^3$ and strong three-dimensional visualization skills. It would be an especially good challenge for students interested in learning the types of linear algebra used in computer graphics.

Combined with the Rigid Folds 2 activity, this provides everything one needs to animate a flat vertex fold opening and closing in Mathematica.

Handout

The handout asks students to fold a simple three-dimensional vertex fold and compute the 3×3 rotation matrices for each crease line. Then, students are asked to multiply them together to see what happens.

The second page can be given separately, if desired, since it gives the conclusion of Question 2 on the previous page. Question 3 asks for an explanation of why the product of the five matrices gave the identity. Question 4 asks for a proof of the general case (which is quite a challenge).

Time commitment

The matrix computations for this activity are tricky to visualize and would take students a full 40–50 minutes to compute and multiply by hand. If a computer algebra system is available, it could take much less time.

Matrices and 3D Origami

Take a square piece of paper and make the below creases to form the 3D **corner of a cube** fold.

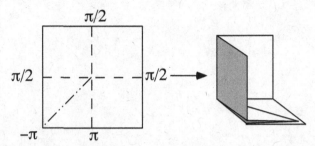

The angles at each crease are the **folding angles**, which is the amount each crease needs to be folded by to make the model.

Question 1: Let χ_i be the 3D, 3×3 rotation matrix that rotates $\mathbb{R}^3$ about the crease line l_i by an angle equal to the folding angle at that crease. Find the five 3×3 matrices $\chi_1, \ldots, \chi_5$ for the above 3D fold. (Assume that the vertex is at the origin and the paper lies in the xy-plane.)

Question 2: What happens when you multiply these matrices together?

Question 3: In the previous question you should have gotten that the product $\chi_1\chi_2\chi_3\chi_4\chi_5 = I$, the identity matrix. Why is this the case?

Be careful with your answer. Remember that the χ_i matrices are rotations about the crease line in the **unfolded** paper.

Question 4: Prove in general that if we are given a 3D single vertex fold with folding matrices $\chi_1, \chi_2, \ldots, \chi_n$, then the product of these matrices, in order, is the identity. Hint: Think of a bug crawling in a circle around the vertex on the folded paper. What rotations would the bug make when it crosses a crease line?

SOLUTION AND PEDAGOGY

This activity is an extension of the Matrix Model for Flat Vertex Folds activity and should only be attempted in a course that has done this previous activity and is studying rotation matrices in $\mathbb{R}^3$. The topic here is three-dimensional origami or, more specifically, *solid angle vertex folds*. This is a specific type of three-dimensional origami where each vertex forms a solid angle in space. In other words, the regions of paper between the creases do not bend or twist—they remain rigid after the model is folded.

While flat folding needs only reflection matrices to model successfully, solid angle vertex folds need rotation matrices in $\mathbb{R}^3$. These rotation matrices χ_i will be determined by the crease line, which will act as the axis of rotation, and the *folding angle* θ_i. The folding angle represents the displacement of the paper from a flat, unfolded position. In other words, the folding angle $\theta_i = \pi -$ the dihedral angle between the planes of paper at the crease line.

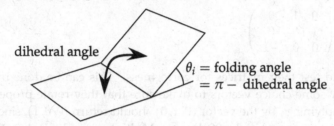

Students need to be familiar with the standard rotation matrices in $\mathbb{R}^3$:

$$R_{yz}(\theta) = \begin{pmatrix} 1 & 0 & 0 \\ 0 & \cos\theta & -\sin\theta \\ 0 & \sin\theta & \cos\theta \end{pmatrix}, \quad R_{xz}(\theta) = \begin{pmatrix} \cos\theta & 0 & -\sin\theta \\ 0 & 1 & 0 \\ \sin\theta & 0 & \cos\theta \end{pmatrix},$$

$$R_{xy}(\theta) = \begin{pmatrix} \cos\theta & -\sin\theta & 0 \\ \sin\theta & \cos\theta & 0 \\ 0 & 0 & 1 \end{pmatrix}.$$

Here $R_{ij}(\theta)$ rotates the ij-plane counterclockwise by angle θ. Label the crease lines $l_1, \ldots, l_5$ starting with the one on the positive x-axis and proceeding counterclockwise. We can compute most of the χ_i matrices by plugging the proper folding angle for θ into one of the above matrices. Care must be taken, however, since the above rotation matrices rotate their respective planes with the assumed orientation that the positive axes are to the right and up. For l_1 this doesn't matter; we get $\chi_1 = R_{yz}(\pi/2)$. But it would be a mistake to think that $\chi_2 = R_{xz}(\pi/2)$ because the crease line l_2 intersects the xz-plane in the wrong orientation, with the positive x- and z-axes in the upper left quadrant. To use the $R_{xz}(\theta)$ matrix, we need to view this rotation from the *other side* of the xz-plane, meaning that our rotation is actually going clockwise, so $\theta = -\pi/2$. That is,

$$\chi_2 = R_{xz}(-\pi/2) = \begin{pmatrix} 0 & 0 & 1 \\ 0 & 1 & 0 \\ -1 & 0 & 0 \end{pmatrix}.$$

The other χ_i matrices for the crease lines that lie on major axes are

$$\chi_1 = \begin{pmatrix} 1 & 0 & 0 \\ 0 & 0 & -1 \\ 0 & 1 & 0 \end{pmatrix}, \chi_3 = \begin{pmatrix} 1 & 0 & 0 \\ 0 & 0 & 1 \\ 0 & -1 & 0 \end{pmatrix}, \chi_5 = \begin{pmatrix} -1 & 0 & 0 \\ 0 & 1 & 0 \\ 0 & 0 & -1 \end{pmatrix}.$$

Crease l_4 is not on one of the main axes, so χ_4 needs to be computed differently. An easy way to find it is via the composition of other rotation matrices; first rotate the crease l_4 to the negative x-axis (call this matrix A), then rotate about the x-axis by the negative of l_4's folding angle (call this matrix B), and then rotate back to the original position (this will be A^{-1}). Then, we have $\chi_4 = A^{-1}BA$. The matrix A requires rotating about the z-axis by $-\pi/4$, so we obtain

$$\chi_4 = \begin{pmatrix} \frac{\sqrt{2}}{2} & -\frac{\sqrt{2}}{2} & 0 \\ \frac{\sqrt{2}}{2} & \frac{\sqrt{2}}{2} & 0 \\ 0 & 0 & 1 \end{pmatrix} \begin{pmatrix} 1 & 0 & 0 \\ 0 & -1 & 0 \\ 0 & 0 & -1 \end{pmatrix} \begin{pmatrix} \frac{\sqrt{2}}{2} & \frac{\sqrt{2}}{2} & 0 \\ -\frac{\sqrt{2}}{2} & \frac{\sqrt{2}}{2} & 0 \\ 0 & 0 & 1 \end{pmatrix}$$

$$= \begin{pmatrix} 0 & 1 & 0 \\ 1 & 0 & 0 \\ 0 & 0 & -1 \end{pmatrix}.$$

Students should test their matrices for correctness. This can be done by multiplying them by some choice vectors to make sure that they rotate properly. For example, multiplying χ_1 by the vector $(0,1,0)$ should return $(0,0,1)$, since this is supposed to rotate the yz-plane by 90 degrees. Multiplying χ_4 by $(-1,0,0)$ should return $(0,-1,0)$.

Multiplying the χ_i matrices together gives

$$\chi_1\chi_2\chi_3\chi_4\chi_5 = I,$$

although they could be multiplied in the reverse order as well and still get the identity. The important thing is to multiply them in order as we travel around the vertex, either clockwise or counterclockwise.

Multiplying five 3×3 matrices can be tedious, and if a computational package that allows matrix multiplication is available, you might want to let your students take advantage of it. On the whole, doing this activity with a mathematical computing package, such as MATLAB, Maple, or Mathematica, at one's side will allow students to explore exactly what their matrices are doing more quickly. Being able to "see" the matrices work can be a great learning experience for students.

Why does this happen?

Intuitively, you and your students might not be surprised that the product of the χ_i matrices gives the identity. After all, the same thing happened for flat vertex folds, right?

But think about what is happening here. We're multiplying matrices that rotate about lines lying in the xy-plane. It's true that the folded paper must come back to where it "started" so as to not rip, but that involves rotating about lines that lie in $\mathbb{R}^3$, no longer in the xy-plane. Why should the mere product of the xy-plane rotations give us the identity?

What is being seen in this activity is an example of a necessary condition for a single vertex to fold up into a three-dimensional shape while keeping all regions of the paper between creases flat and rigid. This necessary condition is not the only one of its kind we could make to model such folds, but it is the most easily-computed, since it only requires xy-plane rotations and doesn't have to keep track of how the paper moves in $\mathbb{R}^3$. Seeing why this always works, however, requires a proof. (Taken from [bel02].)

Theorem: *Let v be a solid angle vertex fold that preserves the flatness of the regions between the creases. Let $\chi_1, \ldots, \chi_n$ be the rotation matrices, in order, about each crease line of v by the respective folding angles. Then $\prod_{i=1}^{n} \chi_i = I$.*

Proof: A classic "bug-walking" argument can help illuminate what's going on. (This will work in the same way as for the flat matrix model but will be more conceptually tricky.) Imagine the unfolded paper sitting in the xy-plane with the vertex v at the origin. Fix the region of the paper, call it F_1, between the creases l_1 and l_n and label the other regions $F_2, F_3, \ldots, F_n$ similarly, going counterclockwise around the vertex. Leaving F_1 fixed in the xy-plane, fold the other regions along the crease pattern into the three-dimensional fold.

Now imagine a bug standing in F_1 on the folded model and let this bug crawl around the vertex in a counterclockwise path (when viewed on the unfolded crease pattern). When the bug crosses crease l_1 it will rotate in space; let L_1 denote the matrix for this rotation. Then the bug will be crawling on region F_2, which no longer lies in the xy-plane. Then it will cross crease l_2; let L_2 denote the matrix for the bug's rotation about this crease line. Continue in this way, defining rotation matrices $L_3, L_4, \ldots, L_n$. Finally, the bug will come back to face F_1 and be in the same orientation as when it began. This implies that

$$L_n L_{n-1} \cdots L_2 L_1 = I.$$

This is the matrix product that most people really have in mind when they think that the result we're trying to prove is "obvious."

Now, what are the L_i matrices? Since F_1 is fixed in the xy-plane, we have $L_1 = \chi_1$. But L_2 is more complicated. One way to envision L_2 is to first unfold l_1, then perform the l_2 crease with matrix χ_2, then refold l_1. The product of these three rotations will result in the bug's rotation around crease l_2 in its three-dimensional position in $\mathbb{R}^3$. That is, we get

$$L_2 = L_1 \chi_2 L_1^{-1}.$$

Similarly, we have $L_3 = L_2 L_1 \chi_3 L_1^{-1} L_2^{-1}$, since we can model the bug's crossing crease l_3 on the folded model by unfolding l_2, then unfolding l_1, then performing χ_3, then refolding l_1 and l_2.

In general, $L_i = $ (redo the previous Ls)χ_i(undo the previous Ls in reverse order). That is,

$$L_i = (L_{i-1} \cdots L_1)\chi_i(L_1^{-1} \cdots L_{i-1}^{-1}).$$

Now, the thing is that these L_i matrices simplify, since they're defined recursively. We get

$$L_1 = \chi_1$$
$$L_2 = \chi_1\chi_2\chi_1^{-1}$$
$$L_3 = (\chi_1\chi_2\chi_1^{-1})(\chi_1)\chi_3(\chi_1^{-1})(\chi_1\chi_2^{-1}\chi_1^{-1}) = \chi_1\chi_2\chi_3\chi_2^{-1}\chi_1^{-1}$$
$$\vdots$$
$$L_i = \chi_1 \cdots \chi_{i-1}\chi_i\chi_{i-1}^{-1} \cdots \chi_1^{-1}.$$

Plugging these into our identity, we get

$$I = L_nL_{n-1} \cdots L_2L_1$$
$$= (\chi_1 \cdots \chi_{n-1}\chi_n\chi_{n-1}^{-1} \cdots \chi_1^{-1})(\chi_1 \cdots \chi_{n-2}\chi_{n-1}\chi_{n-2}^{-1} \cdots \chi_1^{-1}) \cdots (\chi_1\chi_2\chi_1^{-1})(\chi_1)$$
$$= \chi_1\chi_2 \cdots \chi_n.$$

Bingo! □

Alternate proof: One can construct a proof similar to the "rip a region of the paper in half" proof given in the flat vertex case, but it requires paying attention to a few more details.

Each rotation matrix χ_i is determined by two things: the position of the crease line l_i in the xy-plane and the folding angle θ_i. Let F_1 be the region of paper between crease lines l_1 and l_n, and imagine that we rip F_1 into two pieces along a rip from the boundary of the square to the origin. Let F_1' be the ripped region adjacent to l_1 and F_1'' the ripped region adjacent to l_n. Then, we perform the rotation χ_n to F_1'', moving it off the xy-plane. Then we perform χ_{n-1} to this transformed region, and so on, so as to simulate what folding the paper along $l_n, l_{n-1}, \ldots, l_1$ would do to the ripped region F_1''. Since this is a valid solid angle fold, this image of F_1'' should line up with F_1' after all the rotations, giving us

$$\chi_1\chi_2 \cdots \chi_n(F_1'') = I.$$

Now we would like to use the same strategy as in Activity 26 and either employ the three point determination theorem for isometries or the fact that this matrix product is a linear transformation to argue that it must be the identity on all of $\mathbb{R}^3$. But this logic does not immediately follow. For one thing, the three point determination theorem only holds for isometries in a plane. (One would need four points to determine an isometry in $\mathbb{R}^3$.) Similarly, the region F'' is a planar region, and thus only gives us two linearly independent vectors for which we know that the transformation $T = \chi_1\chi_2 \cdots \chi_n$ is the identity, and we would need three to extend this to all of $\mathbb{R}^3$.

What this does tell us, however, is that T is the identity *on all points in the xy-plane*. If we let v be any point in the xy-plane in $\mathbb{R}^3$, then we may write $v = av_1 + bv_2$ where v_1 and v_2 are two linearly independent vectors in F'' in the xy-plane. Then the same argument as done in Activity 26 shows that $T(v) = v$.

From this we just need an argument to show that T is also the identity for points v not in the xy-plane. Since T is the product of rotations, we know that T is an isometry, and thus T is an isometry that leaves the xy-plane fixed. The only isometries of $\mathbb{R}^3$ that leave the xy-plane fixed are the identity and the reflection about the xy-plane.

However, each of the matrices χ_i is a rotation and thus has determinant equal to one. (In fact, they're orthogonal matrices.) The product of a bunch of matrices with determinant one will also have determinant one, and so $\det(T) = 1$. Reflection matrices have determinant -1, and so T cannot be a reflection matrix. Therefore T must be the identity. □

Pedagogy

While the fundamental logic is similar here, the two-dimensional, flat case (as seen in the previous activity) is much more simple. Visualizing reflections in the plane is a lot easier than rotations in three-space. In the two-dimensional case we also have Kawasaki's Theorem to help—there is no easy analog of this for three-dimensional vertex folds.

Working through all the details of this activity may seem very difficult, since there are numerous pitfalls. For example, students are not likely to remember to make sure that they are using the R_{ij} matrices with the proper orientation (with the positive axes in the upper-left quadrant). If technology is available for students to easily check their work, then I think it is entirely reasonable to expect them to figure out all such details. In fact, the first page of the handout makes an excellent test to see if students really understand all that goes on with three-dimensional rotations. If technology is not available, then checking each matrix and multiplying them does become tedious, but the educational value is the same.

While the proof of the general result is very technical, it's only utilizing geometric visualization, careful attention to the order of matrix multiplications, and cancelation of matrices with their inverses. All of this should be doable by a linear algebra student or geometry student with a background in matrices. Instructors can choose to develop the outline of the proof in class and then assign a thorough write-up for homework or to make this proof the subject of a student project. The details of this proof are much better for students to wrestle with and pin down themselves. If such a proof were simply presented in class, the details would likely be wasted on them, resulting in little understanding or growth.

The alternate proof is interesting because it gives an example of the geometric power of the determinant. If students have seen the similar (but easier!) proof for the two-dimensional case, they will be especially eager to try the same approach for 3D folds and in doing so will learn much about the differences and subtleties between isometries in two and three dimensions.

This activity also begins to open the door to some exciting possibilities in drawing three-dimensional folds using computer packages like Maple or Mathematica. However, while the χ_i matrices as presented here could be used to simulate the folded three-dimensional corner in such a computer program, not enough information is included here to, say, animate it opening and closing, for example. Details on how this can be done in some cases will be given in the Rigid Folds 2 activity.

Activity 28
ORIGAMI AND HOMOMORPHISMS

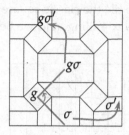

For course: abstract algebra

Summary

We try to prove the following claim: The symmetries of a folded origami model should be able to be determined by the symmetries of its crease pattern. Students prove this using a homomorphism constructed from the origami crease pattern.

Content

This is a fairly advanced activity. For students who have seen homomorphisms in an algebra class, and who have seen and folded some origami models, this activity offers a very hands-on application of homomorphisms to concretely study the symmetries of objects.

Handouts

There are several handouts included in this activity.

(1) Instructions for folding a square twist tessellation. This optional handout gives students an origami model with a lot of symmetry, since it is an origami tessellation.

(2) Introduces notation and asks students to prove that the map φ_σ is a homomorphism.

(3) Asks students to explore examples of the origami homomorphism.

Time commitment

This is a fairly extensive activity. The amount of time required for it will depend greatly on prior class experience with origami and students' familiarity with homomorphisms and symmetry groups. It could require several class periods to explore fully, or only one for a more experienced class.

Folding a Square Twist Tessellation

These instructions show how to tile the classic square twist in square piece of paper. We first look at making a 4 × 4 tiling, and we begin by making a lot of precreases!

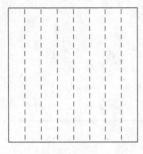

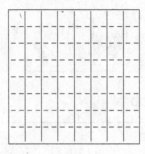

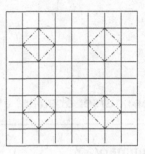

(1) Valley crease the square into 8ths in one direction.

(2) Then valley crease it into 8ths in the other direction.

(3) **Mountain crease** 4 diamond-shaped squares carefully, as shown.

(4) You now have all the creases you need to fold the four square twists. Use the creases shown to the right, where the bold lines are mountains and the dashed lines valleys. Adjacent square twists will rotate in opposite directions. Be persistent!

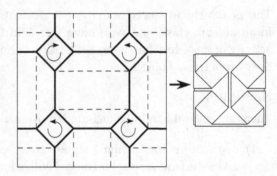

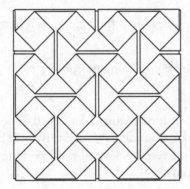

If you succeeded in making a 4 × 4 square twist tessellation, try shooting for an 8 × 8 tessellation! You would need to start by pre-creasing your square into 16ths, and using a larger sheet of paper is recommended.

The Flat-folding Homomorphism

Suppose you have a crease pattern that folds flat. Let γ be a closed, vertex-avoiding curve drawn on the crease pattern that crosses crease lines $l_1, \ldots, l_{2n}$ in order. Let $R(l_i) =$ the transformation that reflects the plane about the line l_i. Since each fold is reflecting part of the paper about the crease, and the paper cannot rip in origami, we have that

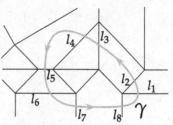

$$R(l_1)R(l_2)R(l_3) \cdots R(l_{2n}) = I,$$

where I is the identity transformation.

Now let σ and σ' be any two faces of the crease pattern C. Define the transformation

$$[\sigma, \sigma'] = R(l_1)R(l_2) \cdots R(l_k),$$

where $l_1, \ldots, l_k$ are the creases, in order, that a vertex-avoiding curve γ crosses going from σ to σ'.

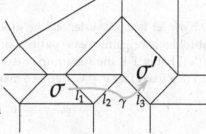

Question 1: Explain why the transformation $[\sigma, \sigma']$ is independent of the choice of the curve γ.

Question 2: Explain why $[\sigma, \sigma''] = [\sigma, \sigma'][\sigma', \sigma'']$ for all faces σ, σ', and σ'' in the crease pattern C.

Question 3: Explain why $[g\sigma, g\sigma'] = g[\sigma, \sigma']g^{-1}$ for all faces $\sigma, \sigma' \in C$ and for any symmetry g of the crease pattern. (In the example shown to the right below, g is a 90° rotation about one of the square twists.)

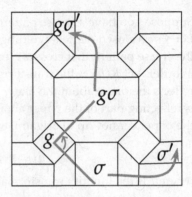

Now let $\text{Isom}(\mathbb{R}^2)$ denote the group of isometries of the plane. Let C be our flat-foldable origami crease pattern, and let $\Gamma \leq \text{Isom}(\mathbb{R}^2)$ be the symmetry group of C. (That is, Γ is the subgroup of isometries that leave C invariant.)

For a fixed face $\sigma \in C$, define a mapping $\varphi_\sigma : \Gamma \to \text{Isom}(\mathbb{R}^2)$ by

$$\varphi_\sigma(g) = [\sigma, g\sigma]g \quad \text{for all } g \in \Gamma.$$

Question 4: Prove that φ_σ is a homomorphism. (That is, prove that $\varphi_\sigma(gh) = \varphi_\sigma(g)\varphi_\sigma(h)$ for all $g, h \in \Gamma$.)

Question 5: Since φ_σ is a homomorphism, what simple fact can we conclude about the image set $\varphi_\sigma(\Gamma)$?

For a fixed face σ of C, we can also define the **folding map** $[\sigma]$ **of** C **toward** σ by

$$[\sigma](x) = [\sigma, \sigma'](x) \text{ for } x \in \sigma' \in C.$$

Question 6: Prove that for any symmetry $g \in \Gamma$, we have that $\varphi_\sigma(g)[\sigma] = [\sigma]g$.

(That is, you want to show these products of transformations are equal, so $\varphi_\sigma(g)[\sigma](x) = [\sigma]g(x)$ for all points x in the crease pattern. Hint: Any point $x \in C$ must lie in a face of the crease pattern, so call this face σ'.)

Question 7: Why does Question 6 imply that $\varphi_\sigma(g) = [\sigma]g[\sigma]^{-1}$ for all $g \in \Gamma$?

What Question 7 says is that the action of any $\varphi_\sigma(g)$ on a flat-folded origami model is equivalent to unfolding it ($[\sigma]^{-1}$), doing an isometry that leaves the crease pattern invariant (g), and then refolding the paper ($[\sigma]$).

Question 8: Explain why this proves that $\varphi_\sigma(\Gamma)$ is the symmetry group of the folded paper!

Finding the Symmetry Group of Origami

Example 1: The classic flapping bird (crane)

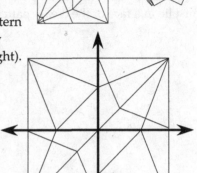

What is the symmetry group Γ of the crease pattern
of the flapping bird? It might be helpful to view
the crease pattern in a set of coordinate axes (right).

    You should get that the symmetry group Γ of this crease pattern has only two
elements. For each of these two elements, call them a and b, determine what $\varphi_\sigma(a)$
and $\varphi_\sigma(b)$ are, for a fixed face σ of the crease pattern.

Conclusion: What does this mean the group $\varphi_\sigma(\Gamma)$ is? Is this the symmetry group
of the folded crane model?
    (Note: You need to think of the folded crane as being flattened into the plane
$\mathbb{R}^2$, not as a 3D model.)

Example 2: The headless crane

Find the symmetry group Γ
of this crease pattern.

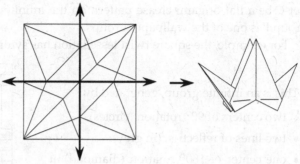

To the right we have labeled a face σ. For each
element $g \in \Gamma$, compute $\varphi_\sigma(g)$, and
thereby determine the group $\varphi_\sigma(\Gamma)$.

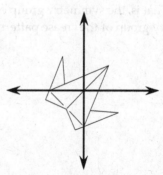

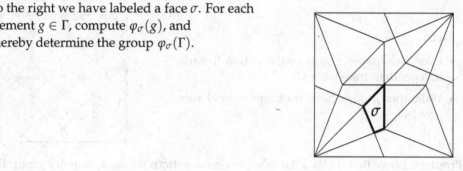

Conclusion: Does your calculation of the group $\varphi_\sigma(\Gamma)$ match the symmetry group
of the folded headless crane? Note that the exact orientation of your headless crane
in the plane under the folding map is determined by our choice of σ.

Example 3: Origami tessellations

Let C be a flat origami crease pattern on the infinite plane $\mathbb{R}^2$ whose symmetry group Γ is one of the wallpaper groups.

For example, the square twist tessellation has symmetry group $\Gamma = \text{p4g}$.

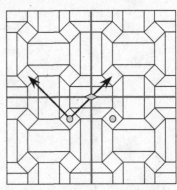

This is an infinite group, generated by

- two centers of 90° rotation (circles),
- two lines of reflection (in grey),
- one center of 180° rotation (diamond) at the intersection of the reflection lines,
- two translation vectors.

Facts:

- Every wallpaper group contains two linearly independent translations.
- Wallpaper groups have no finite normal subgroups.

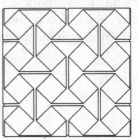

Problem: Prove that if C is a flat origami crease pattern whose symmetry group Γ is a wallpaper group and if the image $\varphi_\sigma(\Gamma)$ is also a wallpaper group, then

$$\varphi_\sigma(\Gamma) \cong \Gamma.$$

That is, the symmetry group of the folded paper will be isomorphic to the symmetry group of the crease pattern.

Follow-up: Can you think of an example of an origami model whose crease pattern is a tessellation but where the folded model is not a wallpaper group? Why does this not contradict the above problem?

SOLUTION AND PEDAGOGY

This is an advanced activity in terms of its use of higher-level mathematics to study origami and is mostly derived from the work in Kawasaki and Yoshida's paper [Kaw88]. It is essential for students to have done some origami prior to this activity, but the purpose of such experience with folding is for the students to have some idea of what paper does when we fold it along a crease pattern. At a minimum have students fold a crane/flapping bird beforehand. (See Activity 20.) Better preparation would be for students to have also folded the classic square twist (see Activity 23) and the square twist tessellation in the first handout.

The real challenge with this activity, however, is not with the origami but with how it forces students to be versatile with abstract algebraic notation *and* make connections between this notation and something real, like the folded piece of paper. This is an incredibly important skill to learn for abstract algebra students. Therefore, this activity offers a chance for students to see an abstract topic like homomorphisms, with all its potentially confusing notation, as telling us something concrete about origami.

Of course, students should have encountered the definition of homomorphisms prior to this activity. And they should have a firm idea of what the symmetry group of an object is (the group of transformations that leave the object invariant).

Handout 1: Folding a square twist tessellation

This handout is optional. It is only provided as a way to give students more experience with folding models with a high degree of symmetry. Origami tessellations are models whose crease pattern is a regular tiling of the plane. Or rather, whose crease pattern could be extended to a tiling of the whole plane, if we had an infinite sheet of paper and the ability to make an infinite number of creases. Compare this to the crease pattern of the crane (flapping bird), with which we could tile in an infinite plane, but then it wouldn't be foldable.

The square twist tessellation is a highly technical fold. That is, the pre-creasing must be made accurately and the creases should be sharp. It is difficult to make the small mountain diagonal creases in step (3). To do so, just pick up the paper, identify the two intersections that make the endpoints of the crease, and pinch the mountain crease into place. Once it has been pinched, it can be made more sharp with one's fingernails.

To make the whole tessellation fold up, it can be very helpful to actually draw the crease pattern on the paper. This is especially true because not all of the creases made during the pre-creasing process are used in the end. Drawing only the creases shown in step (4) on one's paper will aid students and instructors in getting the four square twists to collapse properly and at the same time. The advice, "Be persistent!" is very important.

The handout is highly recommended if instructors want to do Example 3 in Handout 3 with their class. Students will at the very least need to actually see an example of an origami tessellation in order to understand how an origami crease pattern could have its symmetry group be one of the wallpaper groups.

Speaking of which, it is possible to make origami tessellation crease patterns for any wallpaper group desired. Exploring this is well beyond the scope of this activity, and interested readers should consult Eric Gjerde's excellent book [Gje09] and Chris Palmer's seminal work [Pal11] for more details. To whet readers' appetites, the below figure provides one example of a more complicated origami tessellation, designed by the author in 1994. It is based on a $(4, 6, 12)$ Archimedean tiling and is *very* difficult to fold.

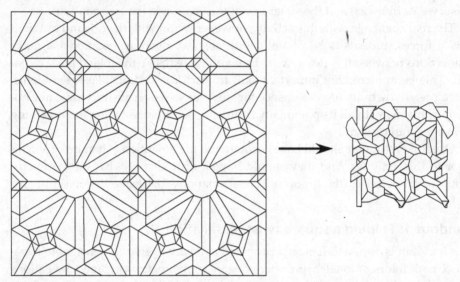

Handout 2: The flat-folding homomorphism

This handout first introduces the notation $R(l_i)$, which is the reflection of the plane about the line containing the crease segment l_i. This is the same notation used in Activity 26 (Matrix Model of Flat Vertex Folds), where the result $R(l_1) \cdots R(l_{2n}) = I$ is explored and proved. In practice, this equation is easy enough to be convincing to students, and so it is merely stated without proof at the beginning on this handout. But classes with enough time should consider exploring Activity 26 first.

The notation $[\sigma, \sigma']$ may need to be described in detail to students first before they sink their teeth into this activity. This notation is taken from [Kaw88], and it is very convenient in terms of capturing what is important, algebraically, with such transformations. It is important for students to know, however, that $[\sigma, \sigma']$ is a product of reflections, and thus is a transformation (an isometry, in fact) of $\mathbb{R}^2$ onto $\mathbb{R}^2$.

Question 1. The fact that $[\sigma, \sigma']$ is independent of our choice of the curve γ is entirely due to the result $R(l_1) \cdots R(l_{2n}) = I$ given at the start of the handout. If γ is a vertex-avoiding curve from σ to σ', then let γ' be another vertex-avoiding curve from σ to σ'. Let $l_1, \ldots, l_{2n}$ be the crease lines that γ crosses and let $l'_1, \ldots, l'_{2k}$ be the crease lines that γ' crosses.

Let γ^{-1} denote the curve that goes backwards along γ. Then the combined curve $\gamma'\gamma^{-1}$ (where we follow γ' and then follow along γ^{-1}) is a closed, vertex-avoiding curve drawn on the crease pattern. Thus we get that

$$R(l'_1) \cdots R(l'_{2k})R(l_{2n}) \cdots R(l_1) = I$$

$$\Rightarrow R(l'_1) \cdots R(l'_{2k}) = R(l_1) \cdots R(l_{2n})$$

and thus $[\sigma, \sigma']$ is the same regardless of whether we used γ or γ'.

Question 2. The fact that $[\sigma, \sigma''] = [\sigma, \sigma'][\sigma', \sigma'']$ follows directly from Question 1. Since our choice of curve γ does not matter, we can perform the transformation $[\sigma, \sigma'']$ by first following a curve from σ to σ' for any other face σ' in the crease pattern and then going from σ' to σ''. That gives us this identity.

Question 3. The figure shown with this question on the handout is meant to be an aid for students to see why $[g\sigma, g\sigma'] = g[\sigma, \sigma']g^{-1}$ for all $\sigma, \sigma' \in C$ and $g \in \Gamma$. In words, this is saying that if I want to fold the creases from $g\sigma$ to $g\sigma'$, where g is a symmetry that leaves the crease pattern invariant, we could instead start at a point in $g\sigma'$, do g^{-1} to get to σ', then reflect about the creases between σ' and σ, and then do g again, literally arriving at the face $g\sigma$. The reason why this works is that since g leaves the crease pattern invariant, the reflections between σ and σ' will be exactly the same as the reflections between $g\sigma$ and $g\sigma'$, only without the transformation g thrown in. Explaining it as simply as that should be plenty for answering this question.

More formally, if we let $l_1, \ldots, l_k$ be the creases that a vertex-avoiding curve γ crosses to get from σ to σ', and if we let gl_i represent the image of the crease l_i under the transformation $g \in \Gamma$, then since g leaves the crease pattern invariant, we have that $R(gl_i) = gR(l_i)g^{-1}$ for all crease lines l_i. Thus

$$[g\sigma, g\sigma'] = R(gl_1)R(gl_2) \cdots R(gl_k) = gR(l_1)g^{-1}gR(l_2)g^{-1} \cdots gR(l_k)g^{-1}$$
$$= gR(l_1)R(l_2) \cdots R(l_k)g^{-1} = g[\sigma, \sigma']g^{-1}.$$

Question 4. To prove that φ_σ is a homomorphism, we use the previous results in the following way:

$$\varphi_\sigma(gh) = [\sigma, gh\sigma]gh = [\sigma, g\sigma][g\sigma, gh\sigma]gh$$
$$= [\sigma, g\sigma]g[\sigma, h\sigma]g^{-1}gh$$
$$= [\sigma, g\sigma]g[\sigma, h\sigma]h = \varphi_\sigma(g)\varphi_\sigma(h).$$

The value in this question is that to show that φ_σ is a homomorphism, one needs to embrace the notation and the abstraction and run with it. Making the connections between the notation and the origami was the job of the previous questions, allowing us to discover the basic properties of the $[\sigma, \sigma']$ transformations. Question 4 requires students to put the interpretation aside for the moment and let the notation do its job. Still, each step in the above proof should "make sense" in terms of the origami, but we don't want students to get too hung up on that. If they understand and believe the solutions to Questions 1–3, then they should be willing to accept the notation used in Question 4. Such trusting and understanding the notation will become especially important in Question 6.

Question 5. Since φ_σ is a homomorphism, we immediately get that the image $\varphi_\sigma(\Gamma)$ is a subgroup of $\mathrm{Isom}(\mathbb{R}^2)$ (see any abstract algebra text, for example, [Gal01]).

Question 6. The notation for the folding map $[\sigma]$ may take some explaining. (And again, we are taking this notation from Kawasaki and Yoshida [Kaw88].) The folding map is literally a function, $[\sigma] : C \to \mathbb{R}^2$, except that writing its domain as the crease pattern C is a little sketchy, since the definition given in the handout only works for points in the interior of the faces of C. Any points on crease lines or at vertices need to be defined a bit differently, and that level of detail is much too tedious to deal with in this activity. (One basically wants $[\sigma]$ to be defined on crease lines and vertices in such a way as to make $[\sigma]$ continuous.)

But since $[\sigma]$ is a function, we can write $[\sigma](x)$ to refer to evaluating the function at a specific point x and write $[\sigma]$ to refer to the function itself, in the same way that we refer to a real-valued function $f(x)$ as just f or as $f(x)$.

Thus, when students see $\varphi_\sigma(g)[\sigma]$ they will likely be confused, as they will be with $[\sigma]g$. This question is a good challenge for digesting abstract notation. Students need to step back and remember that $\varphi_\sigma(g)$ is just a transformation of the plane, and so is $[\sigma]$, and so writing $\varphi_\sigma(g)[\sigma]$ means that we are just composing these two transformations together. Ditto for $[\sigma]g$.

In any case, to prove that $\varphi_\sigma(g)[\sigma] = [\sigma]g$, we try to follow (and manipulate) a point $\varphi_\sigma(g)[\sigma](x)$ to show that it equals $[\sigma]g(x)$. We can do that as follows:

$$
\begin{aligned}
\varphi_\sigma(g)[\sigma](x) &= [\sigma, g\sigma]g[\sigma, \sigma'](x) && \text{where } x \in \sigma' \\
&= [\sigma, g\sigma][g\sigma, g\sigma']g(x) && \text{since } g[\sigma, \sigma'] = [g\sigma, g\sigma']g \text{ by Question 3} \\
&= [\sigma, g\sigma']g(x) && \text{by Question 2} \\
&= [\sigma]g(x) && \text{since } g(x) \in g\sigma'
\end{aligned}
$$

Question 7. This merely follows by right-multiplying both sides of the result of Question 6 by $[\sigma]^{-1}$.

Question 8. This could be the most difficult question in this handout because it requires students to have synthesized the meaning of everything they have learned. Then again, the paragraph before this question connects many of the dots, but in order to get the punchline, a firm understanding of the homomorphism φ_σ and the folding map $[\sigma]$ is required.

The power of the folding map $[\sigma](x)$ is that it represents folding the whole crease pattern. That is, the set of points $[\sigma](C)$ is the folded image of the paper. So $[\sigma]^{-1}$ is the map that unfolds the paper.

Let $h \in \varphi_\sigma(\Gamma)$. We want to show that h is a symmetry of the folded paper. To this end, we have that $h = \varphi_\sigma(g)$ for some symmetry of the crease pattern $g \in \Gamma$. Question 7 tells us that $h = \varphi_\sigma(g) = [\sigma]g[\sigma]^{-1}$.

Writing this in words, we have that the transformation h first unfolds the paper to the original crease pattern ($[\sigma]^{-1}$), then does g, which leaves the crease pattern invariant, and then refolds the paper ($[\sigma]$). In other words, h just unfolds, performs a symmetry, and then refolds the paper, which will leave the folded paper invariant. That is, h is a symmetry of the folded paper and we have that $\varphi_\sigma(\Gamma)$ is the symmetry group of the folded piece of paper.

To do this more formally, consider what the map $\varphi_\sigma(g)$ does to the folded image $[\sigma](C)$ for any $g \in \Gamma$:

$$\varphi_\sigma(g)([\sigma](C)) = [\sigma]g[\sigma]^{-1}([\sigma](C)) = [\sigma]g(C) = [\sigma](C).$$

Here we used the fact that $g(C) = C$ because g leaves the crease pattern C invariant.

Handout 3: Finding the symmetry group of origami

Understanding Handout 2 is greatly facilitated by considering examples. The third handout provides three such examples. The first two, the flapping bird (crane) and the headless flapping bird (headless crane) go very well together because the headless version has a little bit more symmetry than the headed crane, and this is reflected in the homomorphism. The third example concerns an origami tessellation whose symmetry is a wallpaper group.

To compute the actual symmetry groups in Examples 1 and 2, notation will have to be created to describe the transformational symmetries. It will be natural to have one's students use whatever notation was used when exploring, say, the dihedral group D_4, giving the group of symmetries of a square. We will use the notation R_{180} to denote rotation about the origin by $180°$, $R_{y=x}$ to denote reflecting the plane about the line $y = x$, and $R_{y=-x}$ to denote the reflection about the line $y = -x$.

Example 1: The classic flapping bird (crane). We have that the classic crane crease pattern, oriented in the xy-plane as shown on the handout, is only symmetric about the line $y = x$. Thus its group of symmetries is

$$\Gamma = \{I, R_{y=x}\} \cong \mathbb{Z}_2,$$

where I denotes the identity transformation.

Next, we need to see what the image φ_σ is of each element of Γ. This immediately raises a question, which may have come up during your students' exploration of Handout 2, as to how dependent φ_σ is on the choice of the face σ of the crease pattern. Technically, our choice of σ is very important, in that if we were to actually compute formulas or transformation matrices for the symmetries $\varphi_\sigma(g)$, the equations or matrices would change if we changed σ. But the group $\varphi_\sigma(\Gamma)$ won't change. This is because for any two faces $\sigma, \sigma' \in C$, we have that $[\sigma](C) = [\sigma, \sigma'][\sigma'](C)$, which means that the folded images of C with respect to σ and with respect to σ' are congruent by the transformation $[\sigma, \sigma']$ and thus will have isomorphic symmetry groups.

In other words, it doesn't matter what face of the crease pattern we choose for σ, and for the flapping bird we will keep σ arbitrary, while for the headless flapping bird in Example 2 we will need to pick a specific σ.

Getting back to determining $\varphi_\sigma(\Gamma)$, notice that $\varphi_\sigma(I) = [\sigma, \sigma] = I$. Good! The map φ_σ will always map the identity to itself. (We already knew this since φ_σ is a homomorphism, but it's nice that the formula $\varphi_\sigma(g) = [\sigma, g\sigma]g$ verifies it.) For the other element of Γ, we have

$$\varphi_\sigma(R_{y=x}) = [\sigma, R_{y=x}\sigma]R_{y=x} = I.$$

The last equality in the above equation takes some explanation. The idea is that if we took any point $x \in \sigma' \in C$, then $R_{y=x}$ will map x to a face congruent and symmetric to σ' across the line $y = x$, and then the transformation $[\sigma, R_{y=x}\sigma]$ will take the point $R_{y=x}(x)$ back to its original location. This is easier to see of we take $\sigma' = \sigma$, since for $x \in \sigma$ we are literally flipping the face σ across $y = x$ and then folding back to σ. But since this works for the face σ and we are not folding along any other creases besides those in $[\sigma, R_{y=x}\sigma]$, the rest of the paper will follow suit and give us $[\sigma, R_{y=x}\sigma]R_{y=x} = I$ over the entire crease pattern. Note, however, that this phenomenon is quite determined by the actual crease pattern. In particular, we have a crease line right along $y = x$, and this seems to be a big factor in having $R_{y=x}\sigma$ fold back to σ.

We conclude that $\varphi_\sigma(\Gamma) = \{I\}$, which means the folded crane has no symmetries. This is true! Don't let the pleasing simplicity of this classic origami model fool you; it has no rotational symmetry or lines of reflective symmetry when viewed as an object living in $\mathbb{R}^2$. (As a three-dimensional object it does have a plane of reflective symmetry, but since we are modeling our folding process with reflections, we are not keeping track of the different layers of paper at all, and thus we cannot think of the folded image $[\sigma](C)$ as three-dimensional.)

Example 2: The headless crane.

The crease pattern of the headless flapping bird is, of course, almost exactly the same as that of the classic crane/flapping bird. But without the creases for the bird's head, the crease pattern admits more symmetry. Thus we get that the crease pattern's symmetry group is

$$\Gamma = \{I, R_{180}, R_{y=x}, R_{y=-x}\} \cong \mathbb{Z}_2 \times \mathbb{Z}_2.$$

Finding the elements of $\varphi_\sigma(\Gamma)$ is more complicated in this example. In the handout we set σ to be a specific face, shown in the following figures as well. We still get that $\varphi_\sigma(I) = I$ and $\varphi_\sigma(R_{y=x}) = I$ by the same arguments as in Example 1.

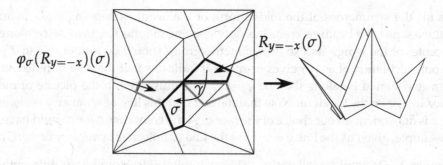

For $\varphi_\sigma(R_{y=-x})$ we do not get the identity, mainly because the line $y = -x$ is not part of the crease pattern! By the formula for φ_σ we get

$$\varphi_\sigma(R_{y=-x}) = [\sigma, R_{y=-x}\sigma]R_{y=-x}.$$

Looking at the above figure, we see that if we perform this transformation to the face σ, then we first reflect it about $y = -x$ to get $R_{y=-x}(\sigma)$, as shown. We then pick the curve γ from this face back to σ as shown and use this to do the transformation $[\sigma, R_{y=-x}\sigma]$. This means we reflect about, in order, the x-axis and then the y-axis (outlined with gray in the figure). This gives us $\varphi_\sigma(R_{y=-x})(\sigma)$ as shown above, and from this we see that, at least as far as the image of σ is concerned,

$$\varphi_\sigma(R_{y=-x}) = R_{y=x}.$$

The same thing would have happened if we had started with any other face of the crease pattern.

For $\varphi_\sigma(R_{180}) = [\sigma, R_{180}\sigma]R_{180}$, we have to follow the face σ again. This time we have to rotate $R_{180}(\sigma)$ first and then follow a curve γ from this face back to σ. This is illustrated in the below figure, where our choice of γ means we'll reflect, in order, about the line $y = x$, the x-axis, and then the y-axis (gray outlines). Seeing the final location of $\varphi_\sigma(R_{180})(\sigma)$ gives us that

$$\varphi_\sigma(R_{180}) = R_{y=x}.$$

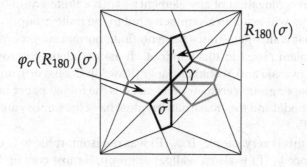

Therefore,

$$\varphi_\sigma(\Gamma) = \{I, R_{y=x}\} \cong \mathbb{Z}_2.$$

This fits the symmetries of the folded form of the headless flapping bird. In an intuitive sense, the headless crane has only one line of reflective symmetry, along the center of the wings, and so its symmetric group should be isomorphic to $\mathbb{Z}_2$. But our calculation of $\varphi_\sigma(\Gamma)$ gives us more; it indicates that this line of reflective symmetry should be along the line $y = x$, as is illustrated in the picture of the folded image on the handout. Note that the fact that this line of symmetry is along $y = x$ is dependent on our choice of the face σ. A different choice for σ could have, for example, given us the line $y = -x$ for the line of reflective symmetry of $[\sigma](C)$.

Example 3: Origami tessellations. This example is structured very differently from the other two. While for the crane and headless crane we could write the symmetry groups Γ and $\varphi_\sigma(\Gamma)$ explicitly, for the infinite square twist tessellation we have that its symmetry group is the wallpaper group (also known as *crystallographic group*) p4g. If students are unfamiliar with the wallpaper groups, they can still explore this example, and it will in fact teach them some things about the wallpaper groups. The illustrations on the handout and description of the generators for p4g are provided to indicate that this symmetry group is infinite, which should make sense to students because obviously this origami crease pattern, viewed on an infinite sheet of paper, will have an infinite number of centers of rotational symmetry and an infinite number of lines of reflection. Each one of these, of course, is generated by combinations of the three generating centers of rotation, the two generating reflection lines, and the two translation vectors. This example offers students a good, geometric example of generators for an infinite group.

The facts presented in the handout for the wallpaper groups are important. In fact, the first fact—that every wallpaper group has two linearly independent translations in its generating set—is what proves the second fact. Our example p4g has plenty of finite subgroups, like $A = \{I, R_{90}, R_{180}, R_{270}\}$ about one of the centers of 90° rotational symmetry. But such a subgroup will not be normal because of the translation vectors. If we call one of these translation vectors $\vec{v}$, then the elements of $A\vec{v}$ will have a different center of rotation than those of $\vec{v}A$, so A isn't normal. The only finite subgroups possible are those generated by the rotations or the reflections; they can't possess any of the translations or the group will become infinite. But then conjugation of any element of such a finite group by one of the translation vectors will result in a symmetry not in the finite group.

This fact—that wallpaper groups have no finite normal subgroups—is important for the problem posed in this handout. It asks students to prove the main theorem from Kawasaki and Yoshida's paper [Kaw88]: If a flat origami crease pattern C has wallpaper group symmetry and so does the folded paper image $[\sigma](C)$, then the folded model and the crease pattern must have the same symmetry group, that is $\varphi_\sigma(\Gamma) \cong \Gamma$.

The proof of this is very simple. If $\varphi_\sigma(\Gamma)$ was not isomorphic to Γ, then $\ker(\varphi_\sigma)$ is not trivial. Since $\varphi_\sigma(\Gamma)$ is also a wallpaper group, it must contain two linearly independent translations, just like Γ. Therefore $\ker(\varphi_\sigma)$ can't contain any translations, and thus it is a finite group. But $\ker(\varphi_\sigma)$ is a normal subgroup of Γ by the first isomorphism theorem, and thus we've found a nonempty finite normal subgroup of our wallpaper group Γ. Since that is impossible, we must have $\varphi_\sigma(\Gamma) \cong \Gamma$.

For the follow-up exercise, to see that the requirement that $\varphi_\sigma(\Gamma)$ is also a wallpaper group is necessary, simply consider the most simple origami tessellation, the infinite square grid:

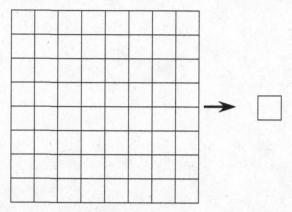

This may seem like an impossible crease pattern to fold, but if you pick one of the squares to be σ, then you can define the folding map $[\sigma](x)$ for any other square in the plane. This is a well-defined map, and thus this is a valid fold, at least as far as imagining infinite paper and an infinite number of creases is concerned. Clearly in this example we have that $\varphi_\sigma(\Gamma) \not\cong \Gamma$, and one can see how the above proof fails in this example. Indeed, here we have that $\ker(\varphi_\sigma) \cong \mathbb{Z} \times \mathbb{Z}$ since the translations in Γ get sent to the identity under φ_σ. So the kernel is able to be a nontrivial normal subgroup of Γ because it is infinite.

Another possible example for the follow-up exercise is the Miura map fold, as seen in Activity 29. In the Miura map fold, folded from an infinite sheet of paper, we have that the folded image is an infinite strip, in which case we would have that $\ker(\varphi_\sigma) \cong \mathbb{Z}$ because only one of the translation vectors is mapped to the identity under φ_σ.

Hopefully by this point readers will be convinced that this activity offers a great opportunity for students to see a very concrete application of symmetry groups, homomorphisms, and the first isomorphism theorem. The fact that these topics are related to origami at all is quite surprising, to say the least!

Activity 29
RIGID FOLDS 1: GAUSSIAN CURVATURE

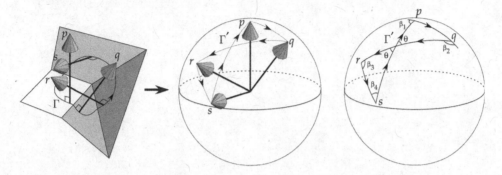

For courses: geometry, differential geometry

Summary

The idea here is to have a sequence of handouts that let students explore the concept of Gaussian curvature, see that paper (and thus all folded models) have zero curvature, and explore what implications this has on rigid origami. Diagrams of the Miura map fold are given to illustrate a model whose vertices pass the rigidity test. Some simple vertex folds and the hyperbolic paraboloid are given as examples that don't work.

Content

This fits right into a differential or topics in geometry class. None of the prior flat folding results are needed to understand this stuff. Several class days would be needed to cover all this (unless, perhaps, the Miura map or hyperbolic paraboloid are given as homework to fold), but this is assuming that Gaussian curvature has not been previously introduced.

Handouts

(1) Introduces Gaussian curvature and lets the students try some easy examples.

(2) Examines the implications of the fact that the Gaussian curvature of a flat sheet of paper is always zero. This leads us to applications to rigid origami.

(3) Instructions for the Miura map fold, a famous example of a rigid fold.

(4) Instructions for the hyperbolic paraboloid, a famous example of a highly non-rigid fold.

Time commitment

The first two handouts can easily take 40 minutes of class time each because of the three-dimensional visualization involved. The two origami instruction sheets will also take time, probably 30 minutes each, but can also be done for homework.

An Introduction to Gaussian Curvature

Definition: The **Gaussian curvature at a point** P on a surface is a real number κ that can be computed as follows: Draw a closed curve Γ on the surface going clockwise around P. Draw unit vectors on the points of Γ that are normal to the surface. Then translate these vectors to the center of a sphere of radius 1 and consider the curve Γ' that they trace on the sphere. (This mapping from Γ to Γ' is called the **Gauss map**.) Then, letting Γ shrink around P, we define the Gaussian curvature at P to be

$$\kappa = \lim_{\Gamma \to P} \frac{\text{Area}(\Gamma')}{\text{Area}(\Gamma)}.$$

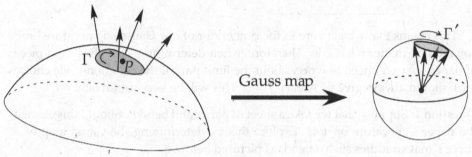

This can be difficult to compute, but not always....

Question 1: What is the Gaussian curvature of a random point on a sphere of radius 1? Radius 2? Radius 1/2?

Question 2: What is the Gaussian curvature of a flat plane?

Question 3: What would happen if you tried to find the Gaussian curvature of a **saddle point**, i.e., the center of a Pringles™ potato chip?

Gaussian Curvature and Origami

In the previous handout, you saw how a flat piece of paper will have zero Gaussian curvature. This is because no matter what our choice of Γ is, the normal vectors along the curve will all be pointing in the same direction, so Area$(\Gamma') = 0$.

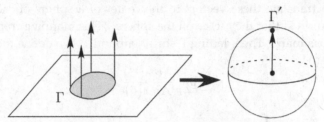

This means that we get zero in the numerator of our Gaussian curvature limit equation no matter what Γ is. Therefore, when determining curvature on a piece of paper, we don't need to worry about the limit part of the equation—one choice for Γ should always give us Area$(\Gamma') = 0$. This will be very useful later on.

Question 1: Suppose that we take a sheet of paper and bend it. Should this change the paper's curvature or not? Explore this by determining the Gauss map of a curve Γ that straddles such a bend, as pictured below.

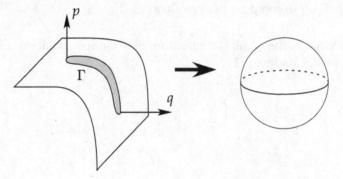

Question 2: Suppose that we make more than one fold, like in an origami model? Draw what the Gauss map should be for the curve Γ shown on the vertex fold below. What should the curvature generated by Γ be? Does this make sense?

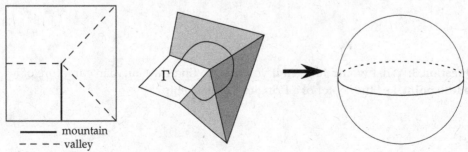

——— mountain
– – – valley

Question 3: The claim that you should have made in Question 2 is this: The Gaussian curvature is zero at every point on a folded piece of paper. Use the Gauss map that you made in Question 2 to prove that this is true for any curve Γ around a 4-valent vertex. (You'll need to use the fact that the area of a triangle on the unit sphere is (the sum of the angles) $- \pi$.)

Question 4: What is the connection between this Gaussian curvature stuff and **rigid origami** (where we pretend that the regions of paper between creases are made of metal and thus are rigid)?

Putting the rigidity criterion to the test

Question 5: Use your conclusions from Question 4 to prove that it is impossible to have a 3-valent folded vertex in a rigid origami model. Draw the Gauss map for such a vertex to back up your argument.

Question 6: Now prove that it is impossible to have a 4-valent vertex in a rigid origami model where **all** of the creases are mountains.

The Miura Map Fold

Japanese astrophysicist Koryo Miura wanted a way to unfold large solar panels in outer space. His fold also makes a great way to fold maps.

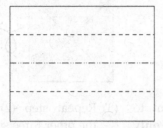

(1) Take a rectangle of paper and mountain-valley-mountain fold it into 1/4ths lengthwise.

(2) Make 1/2 and 1/4 pinch marks on the side (one layer only) as shown.

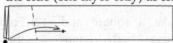

(3) Folding **all layers**, bring the lower left corner to the 1/4 line, as in the picture.

(4) Fold the remainder of the strip behind, making the crease parallel to the previous crease.

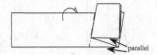

(5) Repeat, but this time use the fold from step (3) as a guide.

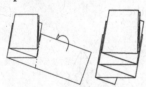

(6) Repeat this process until the strip is all used up. Then **unfold everything**.

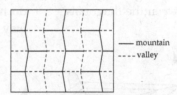

— mountain
---- valley

(7) Now re-collapse the model, but change some of the mountains and valleys. Note how the zigzag creases alternate from all-mountain to all-valley. Use these as a guide as you collapse it...

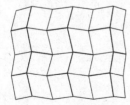

...In the end the paper should fold up neatly as shown to the right. You can then pull apart two opposite corners to easily open and close the model.

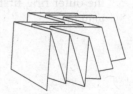

The Hyperbolic Paraboloid

This unusual fold has been rediscovered by numerous people over the years. It resembles a 3D surface that you may recall from Multivariable Calculus.

  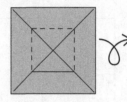

(1) Take a square and crease both diagonals. Turn over.

(2) Fold the bottom to the center, but **only** crease in the middle.

(3) Repeat step (2) on the other three sides. Turn over.

  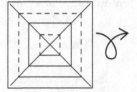

(4) Bring the bottom to the top crease line, creasing **only** between the diagonals.

(5) Then bring the bottom to the nearest crease line. Again, do not crease all the way across.

(6) Repeat steps (4) and (5) on the other three sides. Turn over.

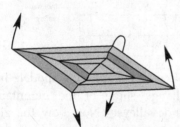 

(7) Now make all the creases at once. It may help to fold the creases on the outer ring first and work your way in.

(8) Once the creases are folded, the paper will twist into this shape, and you're done!

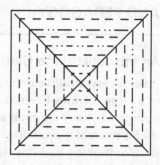

(9) You can make a larger one by folding more divisions in the paper. The key is to have the concentric squares alternate mountain-valley-mountain in the end. You can do steps (1)–(3), do not turn the paper over, then do 1/4 divisions in steps (4)–(6), then turn it over and make 1/8 divisions. Or you could shoot for 1/16ths!

Question: Is the hyperbolic parabola a **rigid origami** model or not? (Could it be made out of rigid sheet metal, with hinges at the creases?) Proof?

SOLUTION AND PEDAGOGY

Handout 1: An Introduction to Gaussian Curvature

This handout gives a very intuitive definition of Gaussian curvature. If your students have seen a more calculus-based definition, then you might want to spend some time describing why they are equivalent. The definition given here turns out to be very useful for modeling rigid origami. The handout does, however, ignore technical details, like proving that this definition is reasonable and well-defined no matter what curve Γ we pick and how we choose to let it shrink to P. But the point of the examples is for students to see that this definition does, indeed, give us a reasonable way of measuring what the curvature of a surface in $\mathbb{R}^3$ might be.

Question 1. The curvature of a sphere of radius 1 will just be 1, since the areas in the numerator and denominator of the limit definition will be equal.

A sphere of radius 2 is more tricky to analyze. One nonrigorous, "arm wavy" approach would be to assert that since a sphere is perfectly symmetric, it should have constant curvature on its surface, and thus the fraction $\text{Area}(\Gamma')/\text{Area}(\Gamma)$ should be constant over all choices of the curve Γ. While this is true, it is not obvious, but it might be the kind of thing an intrepid student with a solid grasp of the concept of curvature would claim. With this assertion, we can take Γ to be something for which $\text{Area}(\Gamma)$ is easy to calculate, like an equator of the sphere. In that case $\text{Area}(\Gamma) = 4\pi 2^2/2 = 8\pi$ and $\text{Area}(\Gamma') = 4\pi 1^2/2 = 2\pi$. Thus $\kappa = 1/4$.

A similar argument will give that the curvature on a sphere of radius $1/2$ should be $\kappa = 4$. In fact, the Gaussian curvature on a sphere of radius r will always be $1/r^2$, and proving this rigorously takes some more work, or at least more knowledge of areas on spheres. For example, suppose that we take Γ to be a perfect circle around our point P on the sphere and we shrink Γ to P evenly, preserving its circle-ness. Then $\text{Area}(\Gamma)$ would be the surface area of the spherical cap with Γ as its boundary. Let r be the radius of the sphere and h be the "height" of the spherical cap made by Γ. (That is, h is the distance from P to the center of the circle Γ inside the sphere.) Then, one can use calculus (either with surfaces of revolution or by looking it up in the back of most calculus books) to get that

$$\text{Area}(\Gamma) = 2\pi rh.$$

Now, under the Gauss map, Γ' will also trace out a circle on the sphere of radius 1. If h' is the height of the spherical cap made by Γ', then we have that $h' = h/r$ since the Γ' cap will be just like the Γ cap with its dimensions scaled down by a factor of r. (That is, the radius r scales down to radius 1, so the height of the cap h will scale down to a height h/r.) Thus,

$$\frac{\text{Area}(\Gamma')}{\text{Area}(\Gamma)} = \frac{2\pi h'}{2\pi rh} = \frac{2\pi h/r}{2\pi rh} = \frac{1}{r^2}.$$

Question 2. A flat plane will, no matter the choice of Γ, have Γ' be a trivial curve—merely a point! Thus Area(Γ') $= 0$ always, and we have $\kappa = 0$. This is a fundamental observation to make for applying Gaussian curvature to origami.

Question 3. This question is a little misleading. A saddle point is an example of a surface having *negative* curvature. The way this happens with our definition is that if we trace a closed curve Γ clockwise around a saddle point P and take the Gauss map, the image Γ' will be traveling *counterclockwise* on the surface of the sphere. Since Γ' is traveling in the opposite direction of Γ, we say that Area(Γ') will be negative, giving us a negative value for κ.

Thus, students will probably find this question confusing. The point is to force them to think about what the Gauss map does for a curve around a saddle point and that the image will have its orientation reversed. If students realize this, you can ask them, "Well, what should going in the opposite direction do to the area?" If they answer, "nothing," then you can argue that this would imply that the curvature at a saddle point can give the same value as the curvature on a sphere. Does that make any sense? So, the convention of making opposite orientations produce negative area gives us a way of distinguishing these different types of surfaces.

Students may think that we're just making this stuff up as we go along, and it's important to tell them that *yes, we are!* The whole idea behind definition-building is to develop notation and concepts that are useful—that allow us to discuss things for which we previously had no language. With the concept of Gaussian curvature, we can describe how much a surface curves by measuring it in a tangible way. And this also gives us a way to classify types of curvature: positive curvature that looks like a sphere, zero curvature that is flat, and negative curvature that looks like a saddle point.

There are many other examples that you could do with your students to help reinforce these ideas. For example:

- What is the curvature of the surface of a cone?

- What is the curvature of a cylinder? (This can be a good preparation for the next handout.)

- What if we measure the curvature of a sphere *from the inside* (like, the bottom half of a sphere, looking at the inside, bowl-shaped region). Will this give us negative curvature or no?

Handout 2: Gaussian Curvature and Origami

The objective of this handout is to take a very elementary observation—that the Gaussian curvature on a flat plane, or piece of paper, is zero everywhere—and use it to make an equally simple, but often confusing, observation about Gaussian curvature on origami. The real motivation is to tie this in to issues of *rigid origami*, where the regions of paper between creases are kept rigid during the folding process.

Question 1. Exploring the Gauss map for a curve Γ that travels over a bent flat surface should indicate that Area(Γ′) = 0, implying that the curvature will be zero. (See the figure below.) However, students who are paying attention to the limit definition of Gaussian curvature can also make the argument that as the curve Γ contracts to a point, it will become so small that we might as well be considering flat, unbent paper again. Either answer is valid for Question 1, but the former one prepares them for the remainder of the handout.

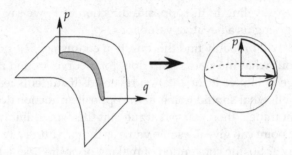

Question 2. Drawing the Gauss map for this curve is quite tricky. All that is really needed is for students to draw their vectors very carefully and pay attention to details, but it's still a challenge to visualize it.

Since the vertex fold given is four-valent, the Gauss map will have only four normal vectors to consider, one for each region of the folded paper. Now, creases are really just bends in the paper, so as Γ crosses a crease line the normal vector will swing from one direction (normal to the previous region of paper) to another direction (normal to the new region being entered by Γ). Thus, we'll have four normal vectors in the Gauss map, and Γ′ will consist of arcs, as in Question 1, connecting the tips of these vectors on the unit sphere. This is shown in the illustration below, where p, q, r, and s are the four vectors normal to the regions of the folded paper.

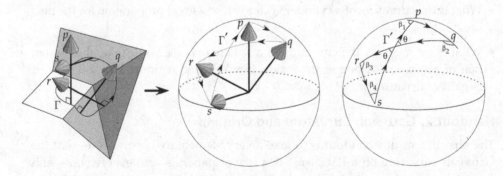

Now, because Γ is a curve on a folded piece of flat paper, we should have Area(Γ′) = 0. One reason for this is because having multiple creases is just compounding the situation in Question 1; if one crease doesn't induce any curvature, then why should more than one crease? Of course, that's a very hand-wavy argument, although it does make intuitive sense. Students, in fact, may find this very persuasive, but they should also see clearly that a more rigorous proof is needed.

In fact, at first it might seem hard to reconcile the fact that we should have Area(Γ') = 0 with the bow-tie spherical polygon that these vectors trace in the Gauss map. But, there are a few things that the students need to convince themselves of before we can do that:

(1) Our crease pattern does generate a bow-tie region in the Gauss map. This will always happen for any four-valent vertex fold that has three valley creases and one mountain crease (or vice versa).

(2) The bow-tie can be thought of as two spherical triangles. If we pay attention to the direction in which Γ and Γ' are traveling, we see that one of these triangles has the same orientation as Γ and the other has opposite orientation.

(3) The angles $\beta_1, \beta_2, \beta_3,$ and β_4 of the bow-tie (shown in the previous illustration) have a strong relationship with the angels, call them α_i, between the crease lines on the paper. If we let α_i be the angle on the region whose normal vector gives us the corner of the bow-tie with angle β_i, then we have $\beta_i = \pi - \alpha_i$.

Item (2) should be reassuring, since the triangle with opposite orientation of Γ will have negative area. Thus our bow-tie spherical polygon might, indeed, have zero area if the two (unoriented) triangles are equal in area.

Item (3) is the most difficult to visualize and leads us to the solution of the next question.

Question 3. The reason why $\beta_i = \pi - \alpha_i$ is because they are supplementary to each other. Focus, for the moment, on the cases of angles β_3 and β_4. No matter how the curve Γ behaves, the normal vectors along it will pivot about a crease line in the same way. In fact, we can think of the normal vectors pivoting around a crease and entering the new region along a trajectory that is perpendicular to the crease line. This trajectory, and the trajectory by which it leaves, will determine the angle β_i. But this means that β_i and α_i will be related as in the below figure, i.e., supplementary to each other.

The cases of β_1 and β_2 are different because the lone mountain crease of the vertex lies between angles α_1 and α_2. If this crease were a valley, then the pivoting normal vectors, say crossing angle α_1's region (vectors s to p to q), would behave as in the other creases, and we'd have the supplementary angle β_1 being in the interior of the spherical polygon made by the Gauss map at vector p. But, since the crease is a mountain, the vector p will swing in the opposite direction (to the right, instead of the left on the previous Gauss map illustration). Angle β_1 is still as before, but since p moved in the other direction, β_1 will be an *external angle* to the spherical polygon (as shown in the illustration). This means that the actual internal angle at p in the spherical bow-tie will be $\pi - \beta_1$. The same thing will happen for the α_2 region of the paper, whose normal vector is q, making its internal angle for the bow-tie $\pi - \beta_2$.

So, if we let θ be the angle at the bow-tie's intersection point t, then we can compute the area contained by our spherical bow-tie, which is Area(Γ'):

$$
\begin{aligned}
\text{Area}(\Gamma') &= (\text{Area of the triangle } srt) - (\text{Area of the triangle } pqt) \\
&= (\beta_3 + \beta_4 + \theta - \pi) - (\pi - \beta_1 + \pi - \beta_2 + \theta - \pi) \\
&= \beta_1 + \beta_2 + \beta_3 + \beta_4 - 2\pi \\
&= \pi - \alpha_1 + \pi - \alpha_2 + \pi - \alpha_3 + \pi - \alpha_4 - 2\pi \\
&= 2\pi - (\alpha_1 + \alpha_2 + \alpha_3 + \alpha_4) = 0.
\end{aligned}
$$

Here we used the fact that the area of a spherical triangle is the sum of the internal angles minus π. (See [Hen01] for details.) This shows how the zero curvature property of the paper is preserved when we fold a four-valent vertex. Higher-valency vertices can be analyzed similarly, although the pictures get much more complicated.

Question 4. This is the first time in these handouts that rigid origami is mentioned, and it's about time! The fact is that when using Gaussian curvature in this way to analyze and model paper folding, we are assuming that the normal vectors are constant over the regions of paper between crease lines. In other words, we are assuming that the paper is rigid except at the creases.

This means that if an origami fold is rigid, then we will be able to compute the Gauss map of any curve Γ drawn on the folded paper and show that we'll have Area(Γ') = 0. This is the answer to expect of students for Question 4.

However, also note that if this Gauss map computation does not work, it will prove that the fold is not rigid. We will use this tool in the next series of questions.

Question 5. Suppose that we have a rigid origami model with a vertex of valency three. Then, if we let our curve Γ be a closed loop around this vertex, we'll get three normal vectors. Because of the creases, none of these vectors will point in the same direction, so the Gauss map will trace out a spherical triangle on the unit sphere. There's no way this can have zero area (unless the triangle is redundant, having one side of length zero, which the three different vectors forbid), so this is impossible to do rigidly.

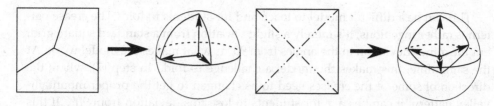

Question 6. This is actually similar to Question 5. If a four-valent vertex has all mountain creases (or, for that matter, all valleys), then the Gauss map will give us a spherical quadrilateral. That is, we won't get the bow-tie phenomenon as in the Question 2 example. Such a spherical quadrilateral cannot have zero area, making this impossible.

Pedagogy. This is definitely an advanced geometry activity, ideally suited for a differential geometry class or a geometry class where spherical geometry is being fully explored. Students need to know the formula for calculating the area of a spherical triangle. They also need to be very comfortable with visualizing normal vectors and how they can move about in three-dimensional space.

Much of this activity is wasted on the students, however, if they don't already have a sense of what it means for origami to be rigid. In particular, they need to understand that while many origami models, like single vertex folds and the Miura map fold, are rigid, there are many that are not. Prior to this activity the students should have the experience of folding a non-rigid origami model. The hyperbolic parabola model serves this purpose, as does the "classic" method of folding the square twist, which can be found in the Folding a Square Twist activity. (Although note that proving the square twist is non-rigid is much harder; see the Rigid Folds 2 activity.)

Handout 3: The Miura Map Fold

This is a very interesting fold that is included here because it might very well be the most famous example of a rigid origami model. It even has applications to space science, of all things.

Koryo Miura invented this fold while searching for a way to collapse a large solar panel into a package that could be attached to a space satellite and fit inside a rocket capsule. This fold seems good for this purpose because one can imagine each parallelogram region of the crease pattern being a solar cell, and these could be taped together to make foldable creases. But the only way in which this would work is if this is a valid rigid origami model (assuming that the solar cells are not flexible). In an attempt to verify that this fold is, indeed, rigid, Miura modeled it using Gaussian curvature, as done in this activity. (See [Miu89].) While this doesn't prove that it's rigid, it does verify that at some level there's nothing preventing it from folding rigidly.

Subsequently, Miura discovered that since this model opens and closes so easily, it makes an ideal map fold. In fact, one can buy Tokyo subway maps that are folded in this way.

This can be a difficult model to teach and for students to fold. The crease pattern is rather ingenious; it's merely a slight variation from a standard square grid, but the minor deviation in the angles from 90° is what makes the model work. At the same time, this makes the model a challenge to fold. In step (7), where the direction of some of the creases need to be changed to get the proper mountain-valley pattern, it can be easy for students to lose this deviation from 90°. If this happens, then when they re-collapse the model they'll lose the zig-zag staggering of folds that steps (2)–(6) produce, and the model won't open and close easily.

One way to keep this from happening is to make sure that the creases made in steps (2)–(6) are *sharp*. Also, calling attention to the angles being made in steps (2)–(6) and emphasizing that these angles must be preserved in the end can help.

The application of this model to solar panel arrays in space satellites can be dramatically made if the instructor prepares a *much larger* Miura map fold in advance. That is, obtain a large rectangle of heavy paper (heavier than copying paper, obtained from an art supply store, say) and in step (1) fold it into 1/8ths or 1/16ths lengthwise. Then proceed as before. In step (7) you'll have more creases to reverse, so do this carefully and with patience. It really pays off; the finished model will be small enough to fit into your pocket but can expand to your entire arm-span (assuming you use large enough paper).

Handout 4: The Hyperbolic Paraboloid

This model has a strange history. Detailed instructions for it can be found in some origami books (like [Jac89]) and on the web, but it has been claimed that this curious folded shape was discovered by Bauhaus artists in 1920s Germany. Numerous origami artists have discovered this model themselves, making it impossible to attribute to any one person.

It's also nothing less than amazing that the paper wants to take on this hyperbolic paraboloid shape when the concentric squares are folded in paper. It can be fun to have students conjecture as to why the paper acts in this way. One way of explaining it is to look at the quarters of the paper divided by the diagonal creases from step (1). In each of these quarters there are parallel creases that alternate mountain-valley-mountain-valley. Now, one sure-fire way of giving a flat sheet strength is to corrugate it. Architects have used this technique for decades, knowing that a vertical, flat concrete slab is not nearly as strong as one which zig-zags back and forth. Our alternating mountain and valley creases in each quadrant of the hyperbolic paraboloid provide such a corrugation, which makes the sides of the paper want to remain straight as the model collapses. The only way to bring the sides of a square together without bending those sides is for two of them to go "up" and two to go "down" in space. That is what we see happening in this origami model.

Teaching tips.   One pitfall when teaching this model is that there will always be students who make their creases in steps (2)–(6) go all the way across the paper. While this isn't a fatal error, it does make the final collapsing more difficult.

The version shown in these diagrams is a 1/4 hyperbolic paraboloid (each quadrant of the paper is folded into fourths). Students should be *strongly* encouraged to make 1/8 or even 1/16 versions. In fact, the instructor should make at least a 1/16 version to show the class, and if possible obtain a large square of paper and attempt a 1/32 version. Such large hyperbolic paraboloid models are *very* impressive. The combination of straight line creases, smooth curves, and geometrical nature of the model is often too much for students; they'll have to make one of their own. The only trick needed for making these larger versions is to keep making the 1/2, 1/4, 1/8, etc. divisions on the same side of the paper, and then flip the paper over before making the last set of divisions. This will make the creases alternate MVMV in the proper way.

Collapsing larger versions is more tricky, however. Once all the creases are made, one needs to start by folding the outer-most square ring, then the next one, and the next, and so on. As this is being done the paper will start to buckle, wanting to take on the hyperbolic paraboloid shape. This can create tension in the paper, making it difficult to fold more of the square rings. One way to overcome this is to collapse the corners, pressing them flat as you work your way to the paper's center. The diagonal creases of the square will be divided into small segments that alternate MVMV as well. But pressing the corners flat is only a way to get at the square rings in the center of the paper; they will need to be opened (relaxed, so as to no longer be pressed flat) for the final model to take on the proper shape.

Answering the question. The main reason for including this model in the activity is because it provides a great example of highly non-rigid origami. Only two regions of the paper (the center-most triangles) remain rigid in this model—all the other trapezoidal regions twist in space. This can be noticed in the actual origami model, but that doesn't explain why it is happening.

It turns out that both of the problems encountered in Questions 5 and 6 of the Gaussian Curvature and Origami handout are present in the hyperbolic paraboloid model. The figure below illustrates this. First of all, when the model is folded, there will be two vertices of valency three on opposite corners of the center-most square. (Note that one diagonal of this center square is not used in the final, folded model.) It is *very* unusual to encounter vertices with exactly three creases in origami, so this by itself is interesting. But the solution to Question 5 implies that the regions bordering these vertices cannot all be rigid.

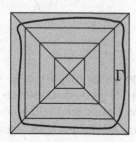
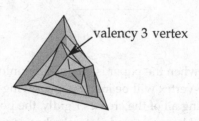

valency 3 vertex

Also, if we draw a curve Γ on the paper that circles through a complete square ring (as shown above), then it will cross only four creases that will either all be mountains or all be valleys. By the same argument used in Question 6, this is impossible if the model were to be rigid. Since such a curve Γ can be drawn on any of these square rings, this proves that non-rigidity will exist throughout the model.

Credits and further problems

As mentioned above, Koryo Miura developed this Gaussian curvature model of rigid origami, apparently sometime in the early 1980s (see [Miu89]). However, David Huffman, of Huffman code fame, explored this same model in a ground-breaking paper in 1976 (see [Huf76]). These discoveries seem to be independent of each other, but as is often the case with researchers in different countries, it is difficult to tell.

In [Miu89], Miura gives another application of Gaussian curvature to a specific origami fold that can make a very interesting homework problem, exam question, or further example for students.

The fold. Take a square piece of paper and fold it in half twice, as one would fold any piece of paper to make it smaller. The crease pattern for this fold is very simple: just four crease lines where the angles between the creases are all 90°. There will be three valley creases and one mountain (or vice versa).

The task. Prove that this four-valent vertex cannot be folded rigidly in a continuous manner from the unfolded state to the flat, folded state, folding all four creases at the same time.

Proof: Since the angles α_i between the creases are all right angles, we also have, using the notation of the previous analysis of four-valent vertices with three valleys and one mountain crease, that $\beta_i = 90°$ for $i = 1, \dots, 4$.

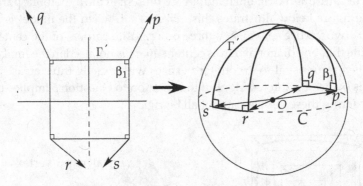

Now, when the paper is completely unfolded, the Gauss map for any curve Γ about the vertex will be just a point, giving us $\mathrm{Area}(\Gamma') = 0$. But, as soon as we start folding all of the creases rigidly, the point in the Gauss map will bloom into a spherical bow-tie quadrilateral whose angles β_i are all right angles. However, it is impossible to have a bow-tie quadrilateral with all right angles drawn on the sphere unless all four of its corners lie along a great circle. (See the previous illustration.) Trying to actually draw such a quadrilateral on, say, an orange or a tennis ball makes this clear; after drawing, say, the two right angles at vectors p and q, whose points on the sphere are connected by a segment of a great circle C, then the sides at p and q that extend perpendicular to C would have to intersect at the "North Pole" of C. Continuing these arcs, the only place where they could meet vectors r and s to form right angles would be on the opposite side of C.

What does this mean? It means that there's no way to go from the unfolded state to a folded state using all four creases at the same time. If we use all four creases, the Gauss map produces a bow-tie that must immediately have the vectors on a great circle.   □

However, the Gauss map does tell us what we can do rigidly with this fold. What does it mean for all the vectors $p, q, r,$ and s to lie on a great circle? We didn't need to mention this before, but the lengths of the arcs between vectors in the Gauss map equals the *folding angle* between these vectors' regions of paper (which is π—the dihedral angle between these planes of paper). So, if p is opposite s on the great circle C, then the folding angle between these regions of paper is π, i.e., their dihedral angle is zero. The p region has been folded flat on top of the s region. Similarly, the q region has been folded flat on top of the r region. In other words, the square of paper has already been folded in half. The lengths of the arcs pq and rs on the Gauss map sphere tell us how much the other creases have been folded.

This tells us a legitimate way to rigidly fold this model: Fold it in half completely and then fold it in half again. Of course, we already knew this, but visualizing what this does to the Gauss map is interesting. Folding it in half the first time keeps, say, vectors p and q on top of each other in the Gauss map (same with r and s). Thus, during this first fold, the Gauss map image is an arc, which gives Area$(\Gamma') = 0$. When the first fold is completely folded flat, we'll have all our vectors on a great circle. Then we can make the second fold, which will split apart p and q (and r and s), giving us the bow-tie in the above illustration, which clearly has Area$(\Gamma') = 0$.

It's remarkable that if we change any of the angles α_i on this fold by a little bit then this argument breaks down completely and the vertex will be rigidly foldable by folding the creases at the same time. In fact, the Miura map fold vertices are only slight deviations from the all-right-angles fold.

Activity 30
RIGID FOLDS 2: SPHERICAL TRIGONOMETRY

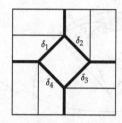

For courses: geometry, differential geometry

Summary

Students use spherical trigonometry to discover strong relationships between the dihedral angles of a four-valent flat vertex fold as it opens and closes rigidly (that is, each region of paper between the creases remains rigid). These results can then be used to prove that certain flat-foldable crease patterns cannot be folded rigidly.

Content

The spherical law of cosines is extensively used, as well as Kawasaki's Theorem (the four-valent case) from the Exploring Flat Vertex Folds activity. This is meant to follow the Rigid Folds 1 activity, although it does not make use of Gaussian curvature. However the results about non-rigid folds fit nicely with the previous non-rigid results. To fully appreciate the results on the square twist's rigidity, students will need to have seen the Folding a Square Twist activity previously.

Also, the dihedral angle relationships are key to making a computer animation (via Maple or Mathematica) of origami folding and unfolding smoothly and rigidly. Combining this with the techniques in the Matrix Model of 3D Vertex Folds activity provides everything that is needed to create such animations.

Handouts and time commitment

The handout has two pages and two parts, which should probably be given separately. The first page helps students discover how certain dihedral angles of a four-valent flat vertex fold are equal and will take about 20–30 minutes to complete fully. The second handout leads students through discovering that some of these dihedral angles will always be greater than others and challenges students to use this to prove rigorously that the square twist is not a rigid fold. That would also take 20–30 minutes, or longer if the students have never seen the square twist before.

Spherical Trigonometry and Rigid Flat Origami 1

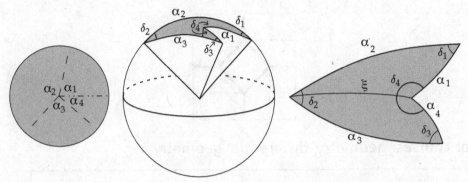

Consider a degree-4 flat vertex fold, as shown above with the angles on the crease pattern $\alpha_1, \ldots, \alpha_4$ and the **dihedral angles** between the regions of folded paper $\delta_1, \ldots, \delta_4$. This is easy to visualize if you imagine the vertex being at the center of a sphere and look at the **spherical polygon** the paper cuts out on the sphere's surface.

If δ_4 is the lone mountain crease, let ζ be an arc on the sphere connecting the δ_4 and the δ_2 corners of this polygon, which divides it into two spherical triangles. Then, we can use the **spherical law of cosines**:

$$\cos \zeta = \cos \alpha_1 \cos \alpha_2 + \sin \alpha_1 \sin \alpha_2 \cos \delta_1, \tag{1}$$

$$\cos \zeta = \cos \alpha_3 \cos \alpha_4 + \sin \alpha_3 \sin \alpha_4 \cos \delta_3. \tag{2}$$

Question 1: Remember that since this vertex folds flat, Kawasaki's Theorem says that $\alpha_3 = \pi - \alpha_1$ and $\alpha_4 = \pi - \alpha_2$. What do you get when you plug these into equation (2) and simplify?

Question 2: Subtract this new equation from equation (1). Use this to find an equation relating the dihedral angles δ_1 and δ_3. What about δ_2 and δ_4?

Spherical Trigonometry and Rigid Flat Origami 2

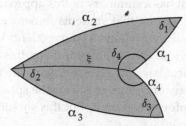

When studying this subject, origami master Robert Lang used spherical trigonometry and the picture above to derive the following equation:

$$\cos \delta_2 = \cos \delta_1 - \frac{\sin^2 \delta_1 \sin \alpha_1 \sin \alpha_2}{1 - \cos \xi}.$$

Question 3: What does this equation tell us about the relationship between the dihedral angles δ_1 and δ_2?

Question 4: Remember that these results assume that the paper is *rigid* between the creases (for otherwise our spherical polygons would not have straight sides). So use your answers to Questions 2 and 3 to prove that the square twist, shown below, **cannot** be folded rigidly. (Bold creases are mountains, non-bold are valleys.)

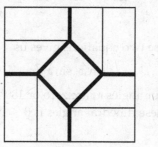

SOLUTION AND PEDAGOGY

This material was originally developed by Robert Lang [Lang01], although it is the same kind of mathematical model that engineers use to study robotic arm movements, what they call *kinematics*. Devin Balkcom's Ph.D. thesis [Bal04] on creating a robot to do simple origami has a summary of this approach.

These handouts do not require that students remember the spherical law of cosines. In fact, it is very unlikely that any of your students will have seen it before. But, since the handout provides the formula for them, this can be thought of as a way to introduce the spherical law of cosines. Students are typically not surprised at all that a version of the law of cosines exists for the sphere, and instructors can refer to [Hen01] for more information on it. But this should not get in the way of the main content of the activity.

Students can have a very hard time visualizing this activity. One must pay careful attention to the difference between the *plane angles* α_i (the angles between the crease lines) and the *dihedral angles* δ_i (the angles between the rigid planes of paper as it's being folded). Some students may need multiple explanations on how one even measures a dihedral angle (by measuring the angle made on a plane orthogonal to the intersection line of the two regions of the paper) and why the dihedral angles will be the same as the interior angles of the spherical quadrilateral (since the fold's vertex is at the center of the sphere, each crease line becomes a radius of the sphere, so a plane tangent to the sphere at one of the angles will be orthogonal to this crease line).

This material does fit in perfectly with three-dimensional solid angle geometry, where similar questions about plane versus dihedral angles are commonplace. See [Cro99] for a fine introduction to such geometry and its relation to polyhedra and Descartes' Theorem.

Question 1

The main thing for students to remember here is that $\cos(\pi - \alpha) = \cos \alpha$ for any angle $0 \leq \alpha \leq \pi$. (So we're actually using the fact, as seen in the Exploring Flat Vertex Folds activity, that all crease angles in a flat vertex fold must be less than 180°.) Then, when we plug the results of Kawasaki's Theorem into equation (2), we get

$$\cos \zeta = \cos \alpha_1 \cos \alpha_2 + \sin \alpha_1 \sin \alpha_2 \cos \delta_3.$$

Question 2

Subtracting these two equations gives us

$$\sin \alpha_1 \sin \alpha_2 (\cos \delta_1 - \cos \delta_3) = 0.$$

Since none of our angles α_i are zero or 180°, this means that $\cos \delta_1 = \cos \delta_3$. Now, the range for these dihedral angles is $0 \leq \delta_1 \leq \pi$ and $0 \leq \delta_3 \leq \pi$, so this implies that $\delta_1 = \delta_3$.

Wow! This means that for a four-valent rigid vertex that folds flat, the opposite crease lines that have the same mountain-valley parity will have equal dihedral angles as the vertex is folded and unfolded.

A similar result is true for δ_2 and δ_4, but since $\delta_4 \geq \pi$, the picture is different. If instead we connect the δ_1 and δ_3 corners of the spherical quadrilateral with an arc ξ, this arc will be outside the quadrilateral since the δ_4 crease is a mountain and thus forms a concave corner of the quadrilateral. See the figure below.

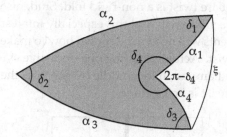

But this still gives us two spherical triangles to which we can apply the spherical law of cosines:

$$\cos \xi = \cos \alpha_2 \cos \alpha_3 + \sin \alpha_2 \sin \alpha_3 \cos \delta_2,$$
$$\cos \xi = \cos \alpha_1 \cos \alpha_4 + \sin \alpha_1 \sin \alpha_4 \cos(2\pi - \delta_4).$$

Using Kawasaki and subtracting then gives us

$$\sin \alpha_1 \sin \alpha_2 (\cos \delta_2 - \cos(2\pi - \delta_4)) = 0.$$

Therefore, we have that $\delta_2 = 2\pi - \delta_4$. In other words, the opposite crease lines that have opposite mountain-valley parity will have equal dihedral angles as well, although since δ_4 is convex it has to be the complement of this angle.

The second page of the handout starts off a bit unfairly—a complicated formula relating the dihedral angles δ_1 and δ_2 is given without proof. The reason for this is that the derivation of this formula is *very* yucky, involving the equations from the previous page, the spherical law of sines, and some horrendous trigonometric manipulations. Students should not be asked to develop this formula by themselves (although it could make for a very hard extra credit problem). Furthermore, the construction of this formula does not help us answer the following questions. For the purposes of studying rigid origami, the emphasis should be placed on what such a formula can tell us and how it can be used.

Question 3

The observation to make here is that the quantity $(\sin^2 \delta_1 \sin \alpha_1 \sin \alpha_2)/(1 - \cos \xi)$ is positive. This is because the $\sin \delta_1$ term is squared and $0 < \alpha_1 < \pi, 0 < \alpha_2 < \pi$, and $\cos \xi \leq 1$. Therefore, we have that $\cos \delta_2 < \cos \delta_1$. Since cosine is a decreasing function from 0 to π, this means that

$$\delta_2 > \delta_1.$$

In other words, when a four-valent flat vertex folds and unfolds rigidly, the opposite-parity pair of opposing dihedral angles will be greater than the equal-parity pair of opposing dihedral angles (where parity means mountain-valley parity).

Question 4

As one example of how these results can be used, students are asked to use them to prove that the classic square twist is a non-rigid fold. Students who have done the Folding a Square Twist activity will find this especially interesting. If that activity has not been done, students will need to be shown how to make this crease pattern fold up into a square twist. Actually folding one does give evidence that the fold is non-rigid; the square diamond in the middle (which does the twisting) does not remain rigid while the twist is being done.

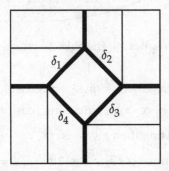

For a proof, suppose that we can fold it rigidly, and consider the figure above where the dihedral angles of the square diamond creases are labeled $\delta_1, \ldots, \delta_4$. Then, notice that at the top vertex, δ_1 is part of a same-parity pair and δ_2 is part of an opposite-parity pair. Thus $\delta_2 > \delta_1$. Then, looking at the right vertex, we see that δ_2 is part of a same-parity pair and δ_3 is part of an opposite-parity pair, so $\delta_3 > \delta_2$. Proceeding then to the bottom vertex and then to the left vertex gives us the chain of inequalities

$$\delta_1 < \delta_2 < \delta_3 < \delta_4 < \delta_1,$$

which is impossible.

Follow-up

If the Folding a Square Twist activity has been done by your students, you should have them go through the other mountain-valley assignments for the square twist to see which of them are rigid. Note, however, that if a dihedral angle contradiction occurs like the one above, then the crease pattern is not rigid. But, if a dihedral angle contradiction does not occur, it does not prove that the fold is rigid, merely that the dihedral angles work out OK. This, combined with the Gaussian curvature model, makes convincing evidence that such crease patterns are indeed rigidly foldable, and it's up to students (and faculty) to decide whether or not these conditions constitute a good enough definition of "rigidity" to qualify as a proof.

Notice also that by taking arccosines of both sides of Lang's equation on the second page of the handout, we obtain a formula for δ_2 in terms of δ_1 and the α_i angles. Thus, we have ways to determine all the dihedral angles of a four-valent flat vertex fold from one angle δ_1. That is, we can think of δ_1 as being a parameter ranging from 0 to π that determines where the rest of the paper should be. In fact, even if we had a larger crease pattern with only four-valent flat-foldable vertices (like the Miura map fold), this one parameter would determine *all* of the other creases' dihedral angles.

Thus, we can determine the folding angles based on what's happening with one crease. This, combined with the matrix transformations of the Matrix Model of 3D Vertex Folds activity, gives us everything that we need to model such folding and unfolding of rigid origami crease patterns in a computer algebra system.

APPENDIX: WHICH ACTIVITIES GO WITH WHICH COURSES?

Presented here is a list of which activities might go best with which mathematics courses. Note, however, that to a certain extent, such a classification is very hard to do. Some activities are relevant to a number of courses. On the other hand, one could argue that *all* of the activities in this book could be looked at as geometry activities or experiments in mathematical modeling. Readers should feel very free to explore these projects themselves to see their relevance.

High-school teachers should also completely disregard this list, as it's designed with the college curriculum in mind. Rather, they should read through these activities and search for things that would be suitable. Of course, many of the activities under Geometry might be good fare for a high-school geometry course, but even here teachers will have to pick and choose because of the differences between college and secondary educational approaches.

Those looking for math club or math circle activities should just read the whole book; literally every activity has potential for groups of math club students to become engaged.

BIBLIOGRAPHY

[Alp00] R. C. Alperin, A mathematical theory of origami constructions and numbers, *New York Journal of Mathematics*, Vol. 6, 2000, 119–133.

[AlpLan09] R. C. Alperin and R. J. Lang, One-, two-, and multi-fold origami axioms, in *Origami⁴: Fourth International Meeting of Origami Science, Mathematics, and Education*, R. J. Lang, ed., A K Peters, Natick, MA, 2009, 371–393.

[Auc95] D. Auckly and J. Cleveland, Totally real origami and impossible paper folding, *The American Mathematical Monthly*, Vol. 102, No. 3, March 1995, 215–226.

[Bal04] D. Balkcom, *Robotic Origami Folding*, Ph.D. thesis, Carnegie-Mellon University Robotics Institute, 2004.

[Bar84] D. Barnette, *Map Coloring, Polyhedra, and the Four Color Problem*, Mathematical Association of America, Washington, D.C., 1984.

[bel02] s.-m. belcastro and T. Hull, Modeling the folding of paper into three dimensions using affine transformations, *Linear Algebra and its Applications*, Vol. 348, 2002, 273–282.

[Bell11] G. Bell, Five intersecting tetrahedra, *Cubism for Fun*, No. 85, 2011, 20–22.

[Belo36] M. P. Beloch, Sul metodo del ripiegamento della carta per la risoluzione dei problemi geometrici (in Italian), *Periodico di Mathematiche, Ser. IV*, Vol. 16, 1936, 104–108.

[Bern96] M. Bern and B. Hayes, The complexity of flat origami, in *Proceedings of the Seventh Annual ACM-SIAM Symposium on Discrete Algorithms*, SIAM, Philadelphia, 1996, 175–183.

[Bon76] J. A. Bondy and U. S. R. Murty, *Graph Theory with Applications*, North Holland, New York, 1976.

[Bri84] D. Brill, Asides: Justin's angle trisection, *British Origami*, No. 107, 1984, 14–15.

[Cox04] D. Cox, *Galois Theory*, John Wiley & Sons, Hoboken, NJ, 2004.

[Cox05] D. Cox, J. Little, and D. O'Shea, *Ideals, Varieties, and Algorithms: An Introduction to Computational Algebraic Geometry and Commutative Algebra*, 2nd ed., Springer, New York, 2005.

[Coxe71] H. S. M. Coxeter, Virus macromolecules and geodesic domes, in *A Spectrum of Mathematics: Essays Presented to H.G. Forder*, J. C. Butcher, editor, Auckland University Press, Auckland, 1971, 98–107.

[Cro99] P. R. Cromwell, *Polyhedra*, Cambridge University Press, Cambridge, UK, 1999.

[Dem99] E. Demaine, M. Demaine, and A. Lubiw, Folding and one straight cut suffice, in *Proceedings of the Tenth Annual ACM-SIAM Symposium on Discrete Algorithms*, SIAM, Philadelphia, 1999, 891–892.

[Dem02] E. Demaine and M. Demaine, Recent results in computational origami, in *Origami^3: Third International Meeting of Origami Science, Mathematics, and Education*, T. Hull, editor, A K Peters, Natick, MA, 2002, 3–16.

[Dem07] E. Demaine and J. O'Rourke, *Geometric Folding Algorithms: Linkages, Origami, Polyhedra*, Cambridge University Press, Cambridge, UK, 2007.

[DiF00] P. Di Francesco, Folding and coloring problems in mathematics and physics, *Bulletin of the American Mathematical Society*, Vol. 37, No. 3, 2000, 251–307.

[Eng89] P. Engel, *Folding the Universe: Origami from Angelfish to Zen*, Vintage Books, New York, 1989.

[Fra99] B. Franco, *Unfolding Mathematics with Unit Origami*, Key Curriculum Press, Emeryville, CA, 1999.

[Fuj82] S. Fujimoto and M. Nishiwaki, *Seizo suru origami asobi no shotai* (Creative Invitation to Playing with Origami, in Japanese), Asahi Culture Center, Tokyo, 1982.

[Fuk89] H. Fukagawa and D. Pedoe, *Japanese Temple Geometry Problems*, Charles Babbage Research Centre, Winnipeg, Canada, 1989.

[Gal01] J.A. Gallian, *Contemporary Abstract Algebra*, 5th ed., Houghton Mifflin Co., Boston, 2001.

[Ger08] R. Geretschläger, *Geometric Origami*, Arbelos, Shipley, UK, 2008.

[Gje09] E. Gjerde, *Origami Tessellations: Awe-Inspiring Geometric Designs*, A K Peters, Wellesley, MA, 2009.

[Gro93] G. M. Gross, *The Art of Origami*, BDD Illustrated Books, New York, 1993.

[Haga95] K. Haga, Origamics, Parts 1–4 (in Japanese), *ORU*, No. 9–12, Summer 1995–Spring 1996, 64–67, 68–72, 60–64, and 60–64, respectively.

[Haga99] K. Haga, *Origamics: Fold a Square Piece of Paper and Make Geometrical Figures, Part 1* (in Japanese), Nihon-hyouron-sha, Tokyo, 1999.

[Haga02] K. Haga, Fold paper and enjoy math: origamics, in *Origami^3: Third International Meeting of Origami Science, Math, and Education*, T. Hull, editor, A K Peters, Natick, MA, 2002, 307–328.

[Haga08] K. Haga, *Origamics: Mathematical Explorations Through Paper Folding*, World Scientific Publishing Co., River Edge, NJ, 2008.

[Hart01] G. W. Hart and H. Picciotto, *Zome Geometry: Hands-on Learning with Zome Models*, Key Curriculum Press, Emeryville, CA, 2001.

[Hat05] K. Hatori, How to divide the side of square paper, available at http://www.origami.gr.jp/People/CAGE_/divide/index-e.html.

[Hen01] D. Henderson, *Experiencing Geometry in Euclidean, Spherical, and Hyperbolic Spaces*, 2nd ed., Prentice Hall, Upper Saddle River, NJ, 2001.

[Hil97] P. Hilton, D. Holton, and J. Pedersen, *Mathematical Reflections: In a Room with Many Mirrors*, Springer, New York, 1997.

[Huf76] D.A. Huffman, Curvature and creases: a primer on paper, *IEEE Transactions on Computers*, Vol. C-25, No. 10, Oct. 1976, 1010–1019.

[Hull94] T. Hull, On the mathematics of flat origamis, *Congressus Numerantium*, Vol. 100, 1994, 215–224.

[Hull02-1] T. Hull, The combinatorics of flat folds: a survey, in *Origami^3: Third International Meeting of Origami Science, Mathematics, and Education*, T. Hull, editor, A K Peters, Natick, MA, 2002, 29–38.

[Hull02-2] T. Hull, editor, *Origami^3: Third International Meeting of Origami Science, Mathematics, and Education*, A K Peters, Natick, MA, 2002.

[Hull03] T. Hull, Counting mountain-valley assignments for flat folds, *Ars Combinatoria*, Vol. 67, 2003, 175–188.

[Hull05-1] T. Hull, Origametry part 6: basic origami operations, *Origami Tanteidan Magazine*, No. 90, March 2005, 14–15.

[Hull05-2] T. Hull, Exploring and 3-edge-coloring spherical buckyballs, unpublished manuscript, 2005.

[Hull11] T. Hull, Solving cubics with creases: the work of Beloch and Lill, *American Mathematical Monthly*, Vol. 118, No. 4, 2011, 307–315.

[HullCha11] T. Hull and E. Chang, The flat vertex fold sequences, in *Origami^5: Fifth International Meeting of Origami Science, Mathematics, and Education*, A K Peters/CRC Press, Natick, MA, 2011, 599–607.

[Hus79] K. Hushimi and M. Hushimi, *Origami no kikagaku* (Geometry of Origami, in Japanese), Nihon-hyoron-sha, Tokyo, 1979.

[Hus80] K. Hushimi, Trisection of angle by H. Abe in Origami no kagaku (The Science of Origami, in Japanese), *Saiensu* (the Japanese edition of *Scientific American*), Oct. 1980 (appendix in separate volume), 8.

[Huz92] H. Huzita, Understanding geometry through origami axioms: is it the most adequate method for blind children?, in *Proceedings of the First International Conference on Origami in Education and Therapy*, J. Smith, editor, British Origami Society, London, 1992, 37–70.

[Huz89] H. Huzita and B. Scimemi, The algebra of paper-folding (origami), in *Procedings of the First International Meeting of Origami Science and Technology*, H. Huzita, editor, Ferrara, Italy, 1989, 205–222.

[Ike09] U. Ikegami, Fractal crease patterns, in *Origami^4: Fourth International Meeting of Origami Science, Mathematics, and Education*, R. J. Lang, ed., A K Peters, Natick, MA, 2009, 31–40.

[Jac89] P. Jackson, *The Complete Origami Course*, W. H. Smith, New York, 1989.

[Jus84] J. Justin, Coniques et pliages (in French), *PLOT, APMEP Poitiers*, No. 27, 1984, 11–14.

[Jus86] J. Justin, Mathematics of origami, part 9, *British Origami*, No. 118, 1986, 28–30.

[Jus97] J. Justin, Toward a mathematical theory of origami, in *Origami Science and Art: Proceedings of the Second International Meeting of Origami Science and Scientific Origami*, K. Miura, editor, Seian University of Art and Design, Otsu, 1997, 15–29.

[Kas83] K. Kasahara and J. Maekawa, *Viva! Origami*, Sanrio, Tokyo, 1983.

[Kas87] K. Kasahara and T. Takahama, *Origami for the Connoisseur*, Japan Publications, New York, 1987.

[Kaw88] T. Kawasaki and M. Yoshida, Crystallographic flat origamis, *Memoirs of the Faculty of Science, Kyushu University, Ser. A*, Vol. 42, No. 2, 1988, 153–157.

[Koe68] J. Koehler, Folding a strip of stamps, *Journal of Combinatorial Theory*, Vol. 5, 1968, 135–152.

[Lang95] R. J. Lang, *Origami Insects and Their Kin*, Dover, New York, 1995.

[Lang01] R. J. Lang, Tessellations and twists, unpublished manuscript, 2001.

[Lang03] R. J. Lang, Origami and geometric constructions, unpublished manuscript, 2003.

[Lang04-1] R. J. Lang, ReferenceFinder, available at http://www.langorigami.com/science/reffinder/reffinder.php4.

[Lang04-2] R. J. Lang, Angle quintisection, available at http://www.langorigami.com/science/quintisection/quintisection.php4.

[Lang09] R. J. Lang, editor, *Origami⁴: Fourth International Meeting of Origami Science, Mathematics, and Education*, A K Peters, Natick, MA, 2009.

[Lang11] R. J. Lang, *Origami Design Secrets: Mathematical Methods for an Ancient Art*, 2nd ed., A K Peters/CRC Press, Boca Raton, 2011.

[Law89] J. Lawrence and J. E. Spingam, An intrinsic characterization of foldings of euclidean space, *Annales de l'institut Henri Poincaré (C) Analyse non linéaire*, Vol. 6, 1989 (supplement), 365–383.

[Lill1867] E. Lill, Résolution graphique des equations numériques d'un degré quelconque à une inconnue (in French), *Nouv. Annales Math.*, Ser. 2, Vol. 6, 1867, 359–362.

[Lot1907] A. J. Lotka, Construction of conic sections by paper-folding, *School Science and Mathematics*, Vol. 7, No. 7, 1907, 595–597.

[Lun68] W. F. Lunnon, A map-folding problem, *Mathematics of Computation*, Vol. 22, No. 101, 1968, 193–199.

[Mae02] J. Maekawa, The definition of iso-area folding, in *Origami³: Third International Meeting of Origami Science, Mathematics, and Education*, T. Hull, editor, A K Peters, Natick, MA, 2002, 53–59.

[Maor98] E. Maor, *Trigonometric Delights*, Princeton University Press, Princeton, NJ, 1998.

[Mar98] G. E. Martin, *Geometric Constructions*, Springer, New York, 1998.

[Mes86] P. Messer, Problem 1054, *Crux Mathematicorum*, Vol. 12, No. 10, 1986, 284–285.

[Miu89] K. Miura, A note on intrinsic geometry of origami, in *Proceedings of the First International Meeting of Origami Science and Technology*, H. Huzita, editor, Ferrara, Italy, 1989, 239–249.

[Mon79] J. Montroll, *Origami for the Enthusiast*, Dover, New York, 1979.

[Mon09] J. Montroll, *Origami Polyhedra Design*, A K Peters, Natick, MA 2009.

[Mor24] F. V. Morley, Discussions: a note on knots, *American Mathematical Monthly*, Vol. 31, No. 5, 1924, 237–239.

[Mos] J. Mosely, *The Menger Sponge* and *The Business Card Menger Sponge Project*, available at http://theiff.org/oexhibits/menger02.html.

[ORo11] J. O'Rourke, *How to Fold It: The Mathematics of Linkages, Origami, and Polyhedra*, Cambridge University Press, Cambridge, UK, 2011.

[Ow86] F. Ow, Modular origami (60° unit), *British Origami*, No. 121, 1986, 30–33.

[Pal11] C. Palmer and J. Rutzky, *Shadowfolds: Surprisingly Easy-to-Make Geometric Designs in Fabric*, Kodansha International, New York, 2011.

[Pet01] I. Peterson, *Fragments of Infinity*, John Wiley & Sons, New York, 2001.

[Riaz62] M. Riaz, Geometric solutions of algebraic equations, *American Mathematical Monthly*, Vol. 69, No. 7, 1962, 654–658.

[Rob77] S. A. Robertson, Isometric folding of Riemannian manifolds, *Proceedings of the Royal Society of Edinburgh*, Vol. 79, No. 3–4, 1977–78, 275–284.

[Robi00] N. Robinson, Top ten favorite models, *British Origami*, No. 200, Feb. 2000, 1, 34–42.

[Row66] T. S. Row, *Geometric Exercises in Paper Folding*, Dover, New York, 1966. (This is a reprint of older editions dating back to 1893.)

[Rupp24] C. A. Rupp, On a transformation by paper folding, *American Mathematical Monthly*, Vol. 31, No. 9, 1924, 432–435.

[Sch96] D. P. Scher, Folded paper, dynamic geometry, and proof: a three-tier approach to the conics, *Mathematics Teacher*, Vol. 89, No. 3, 1996, 188–193.

[Ser03] L. D. Servi, Nested square roots of 2, *American Mathematical Monthly*, Vol. 110, No. 4, 2003, 326–330.

[Smi03] S. Smith, Paper folding and conic sections, *Mathematics Teacher*, Vol. 96, No. 3, 2003, 202–207.

[Tan01] J. Tanton, A dozen questions about the powers of two, *Math Horizons*, Sept. 2001, 5–10.

[Tuc02] A. Tucker, *Applied Combinatorics*, 4th ed., John Wiley & Sons, New York, 2002.

[Wang11] P. Wang-Iverson, R. J. Lang, M. Yim, *Origami^5: Fifth International Meeting of Origami Science, Mathematics, and Education*, A K Peters/CRC Press, Natick, MA, 2011.

[Wei1] E. W. Weisstein, Cubohemioctahedron, *MathWorld—A Wolfram Web Resource*, available at http://mathworld.wolfram.com/Cubohemioctahedron.html.

[Wei2] E. W. Weisstein. Chiral, *MathWorld—A Wolfram Web Resource*, available at http://mathworld.wolfram.com/Chiral.html.

[Wen74] M. J. Wenninger, *Polyhedron Models*, Cambridge University Press, London, 1974.

Index

Printed in the United States
by Baker & Taylor Publisher Services

Printed in the United States
by Baker & Taylor Publisher Services